INTERNATIONAL ENERGY AGENCY

D1784526

In support of the G8 Plan of Action

OPTIMISING RUSSIAN NATURAL GAS

Reform and Climate Policy

INTERNATIONAL ENERGY AGENCY

The International Energy Agency (IEA) is an autonomous body which was established in November 1974 within the framework of the Organisation for Economic Co-operation and Development (OECD) to implement an international energy programme.

It carries out a comprehensive programme of energy co-operation among twenty-six of the OECD's thirty member countries. The basic aims of the IEA are:

- To maintain and improve systems for coping with oil supply disruptions.
- To promote rational energy policies in a global context through co-operative relations with non-member countries, industry and international organisations.
- To operate a permanent information system on the international oil market.
- To improve the world's energy supply and demand structure by developing alternative energy sources and increasing the efficiency of energy use.
- To assist in the integration of environmental and energy policies.

The IEA member countries are: Australia, Austria, Belgium, Canada, the Czech Republic, Denmark, Finland, France, Germany, Greece, Hungary, Ireland, Italy, Japan, the Republic of Korea, Luxembourg, the Netherlands, New Zealand, Norway, Portugal, Spain, Sweden, Switzerland, Turkey, the United Kingdom and the United States. The European Commission takes part in the work of the IEA.

ORGANISATION FOR ECONOMIC CO-OPERATION AND DEVELOPMENT

The OECD is a unique forum where the governments of thirty democracies work together to address the economic, social and environmental challenges of globalisation. The OECD is also at the forefront of efforts to understand and to help governments respond to new developments and concerns, such as corporate governance, the information economy and the challenges of an ageing population. The Organisation provides a setting where governments can compare policy experiences, seek answers to common problems, identify good practice and work to co-ordinate domestic and international policies.

The OECD member countries are: Australia, Austria, Belgium, Canada, the Czech Republic, Denmark, Finland, France, Germany, Greece, Hungary, Iceland, Ireland, Italy, Japan, Korea, Luxembourg, Mexico, the Netherlands, New Zealand, Norway, Poland, Portugal, the Slovak Republic, Spain, Sweden, Switzerland, Turkey, the United Kingdom and the United States. The European Commission takes part in the work of the OECD.

This book is dedicated to Gordon Duffus, who died on 28 February 2006.
He joined the IEA as Head of Division for the Office of Non-member Countries in 2002.
Gordon's intellect, caring leadership and positive outlook on life continues to be an inspiration to us all.

FOREWORD

The timing of this book could not be more opportune. Russia is the world's largest gas producer and exporter. Its role in the emerging global gas market will only gain in importance, as growth is projected both in Russian domestic demand and in international requirements. Russia has been a dependable supplier over the past decades but short-lived interruptions in early 2006 due to price disputes with Ukraine and extreme cold weather conditions have raised concerns about its future reliability. As the country's key producing fields decline, Gazprom's ability to increase gas production is critical to international energy security. In this context, the IEA is especially interested in enhancing a dialogue with the Russian government and Gazprom on increasing transparency in the gas sector and promoting the needed energy sector reforms that would contribute to more efficient pricing and sustainable production of natural gas in Russia.

As a Party to the Kyoto Protocol, Russia could find an additional incentive for reform in the synergies available between its climate policy goals and the natural gas sector. This study examines the potential to reduce GHG emissions in the Russian natural gas sector and to limit natural gas flaring by oil companies. The IEA estimates that at least 30 billion cubic meters – a fifth of the country's exports to European OECD countries – could be saved annually by the introduction of more advanced, available technology and the implementation of energy efficiency. Such investments would be all the more attractive as they would also generate reductions equivalent to 150 million tonnes of CO_2 equivalent, that could also be sold on the emerging carbon markets. Russia's ability to identify concrete projects that deliver greenhouse gas reductions would furthermore be attractive to OECD countries seeking carbon trading opportunities.

Much still needs to be done in Russia to take advantage of these opportunities and to translate their potential into commercial transactions. We hope that this study will focus attention on these key energy policy needs, and foster a dialogue among Russian stakeholders (government, domestic and international investors, Gazprom and gas consumers across Eurasia). In the spirit of the G8 Gleneagles Summit and in line with the focus of the G8 Summit in St. Petersburg, progress here would be a significant contribution to global energy security, economic growth and a cleaner environment.

Claude Mandil
Executive Director

ACKNOWLEDGEMENTS

This book could not have been completed without the insight and input provided by the Russian government, non-government and industry officials with whom the authors met over the course of last year. Special appreciation of the authors goes to the Ministry of Industry and Energy, the Ministry of Economic Development and Trade, OAO "Gazprom", Centergazservice-opt, Institute of Geopolitics and Energy Security of Russia, National Carbon Union, Russian Regional Environmental Centre, World Wildlife Fund, and numerous independent experts working on climate policy in Russia. The IEA also thanks the United States National Oceanic and Atmospheric Administration (NOAA), the Global Gas Flaring Reduction Private-Public Partnership team of the World Bank and the Russia-EU Technology Centre for its round table meeting on gas flaring.

The principal authors of this book are Isabel Murray and Alexandrina Platonova-Oquab, of the Office of Non-member Countries. Martina Bosi launched this project during her stay at the IEA, and played an active role in the initial drafting. Gordon Duffus and Richard Bradley provided supervision and encouragement. This book also benefitted from the input of several other IEA colleagues. Special thanks to Richard Baron for his considerable input on climate policy issues, and also to Cédric Philibert for his constructive comments. Christian Besson of Schlumberger provided essential support with his useful insights and input, as well as his colleague Tom O'Gallagher. The Energy Statistics Division also provided invaluable support, especially Roberta Quadrelli, Stève Gervais and Riccardo Quercioli.

Other IEA Secretariat staff without whom this book could not have been completed include Meredydd Evans, who provided insights on energy-efficiency issues, Rebecca Gaghen and Sylvie Stephan, who provided guidance on the key messages, and Jenny Gell, who gave valuable assistance in editing the book. We also thank Bertrand Sadin, who prepared the maps and figures, Corinne Hayworth, who designed the cover, and Muriel Custodio, who supervised the production stage.

We would like to acknowledge the financial contributions of the governments of Canada and Norway.

For any questions or comments please contact the Office of Non-member Countries:

Alexandrina Platonova-Oquab
Tel: 33-1-40-57-66-61
Fax: 33-1-40-57-65-79
e-mail: alexandrina.platonova@iea.org

Isabel Murray
Tel: 33-1-40-57-65-89
Fax: 33-1-40-57-65-79
e-mail: isabel.murray@iea.org

TABLE OF CONTENTS

ANNEXES

LIST OF MAPS

LIST OF FIGURES

LIST OF TABLES

LIST OF BOXES

EXECUTIVE SUMMARY

How can Russia take advantage of its huge energy-efficiency potential to enhance domestic and global energy security? How can Russia build on the synergies between energy sector reform and climate policy? What could Russia contribute to help achieve international Kyoto Protocol commitments? What can the climate community expect from Russia's participation in the Kyoto Protocol?

With these questions in mind, this book assesses the potential of reducing leaks, technical losses and ultimately greenhouse gas (GHG) emissions in Russia's gas transmission and distribution sector, as well as the prospects for reducing gas flaring. This work focusses on energy security and reduction of emissions.[1] It identifies the barriers to achieving these critical objectives responsible for 15% of Russia's GHG emissions. It points to the structural and regulatory reform needs to ensure the sustainable functioning of Russia's gas sector as well as the effective implementation of Kyoto flexibility mechanisms to reduce the sector's GHG emissions.

The lack of investment in maintenance and refurbishment in Russian gas infrastructure suggests a large potential for GHG emission reductions. This will be attractive for any country seeking to use "flexibility mechanisms" to meet their emissions reduction targets. Achieving real GHG emission reductions through projects is more rewarding to buyers, as it demonstrates the environmental benefit attached to the GHG transaction.

OVERVIEW OF RUSSIA'S NATURAL GAS SECTOR

The era of relatively cheap Gazprom gas ending

Russia's proven natural gas reserves amount to 47 trillion cubic meters, 26% of the world's total. In 2004, Gazprom held licenses to fields accounting for 60% of these reserves; 21% is held by other producers, with the remaining 19% unallocated. The bulk of Russian gas production comes from three super-giant fields which are now in decline at a rate of 20 bcm/year. Gazprom is facing a steep rise in production costs as it must develop new fields in deeper strata and/or in the Arctic and other difficult-to-develop regions to compensate for the depletion at current fields, let alone to increase production in line with its production targets. Gazprom annual investments have been in the order of USD 7 billion since 2003 but much has been directed at foreign acquisitions and new export infrastructure. In 2005, Gazprom's management approved a more than

---

1. Demand-side efficiency improvements are not in the scope of this publication. See *Coming in from the Cold* (IEA, 2004a) for a discussion of efficiency measures in Russia's district heating systems in particular.

40% increase in its investment programme to USD 10.8 billion, much of the increase being directed to the North European Gas Pipeline project. This corresponds to the IEA's estimate in its *World Energy Investment Outlook* (IEA, 2003) of USD 11 billion per year required to bring on new sources of Russian gas and to upgrade and maintain gas infrastructure. The IEA is concerned about the priority Gazprom seems to be placing on foreign acquisitions and export infrastructure as opposed to its domestic network and upstream.

Lack of competition in Russia's upstream gas sector

For future gas capacity, Gazprom tends to focus on mega-projects with demanding engineering requirements and the concurrent "mega" investment needs. Tapping such expensive reserves is unlikely to result in cheap gas. The increasing number of non-Gazprom gas producers of both associated and non-associated gas represents a huge potential for efficiency gains from more competition in Russia's upstream gas sector. A growing number of non-Gazprom gas producers and foreign investors are interested in providing substantial capital if there was a reliable and transparent access to the gas transportation network and gas-processing capacity controlled by Gazprom. Absent such conditions, significant volumes of gas associated with oil extraction are still being flared. The Russian Energy Strategy, approved in August 2003, projects non-Gazprom production accounting for 20% of total Russian production in 2020.

The risks inherent in Gazprom's Central Asian strategy

Gazprom has also focussed its efforts on Central Asian gas reserves since 2003, as opposed to development of its own or that of other Russian gas producers. More disconcerting is the apparent lack of investments over the past years in either the upstream or the pipeline infrastructure in Central Asia. Investments seem even less likely since 2006 as tense negotiations over gas prices continue – both for Russian and for Ukrainian imports from Turkmenistan. Obtaining market prices seems, quite understandably, to be a key objective in Gazprom's commercial relationship with foreign customers. The same objective should be set for its domestic market.

Energy-efficiency improvements can reduce pressure on gas deliverability

With the era of "cheap" gas over, and an uncertain relationship with Turkmenistan ahead, Gazprom is facing major choices. A clear win-win option to reduce pressure on gas deliverability is a strategy to slow rising domestic gas demand as the Russian economy grows, through intensifying energy-efficiency programmes and more market-based gas pricing. Energy-efficiency targets have been the centrepiece of Russia's Energy Strategy over the past decade, yet the low domestic gas prices and lack of metering equipment have stymied efforts to improve energy efficiency.

The IEA is concerned that these factors will begin to affect Russia's position as a secure and reliable supplier. The Energy Strategy shows the country's awareness of its energy-efficiency potential. It could use this potential to slow demand growth and help manage the above problems.

The synergies between energy efficiency and climate policy

The synergies between a more efficient use of gas resources and GHG emission reductions are clear in Russia, and could be exploited through the Kyoto Protocol mechanisms. This study examines the potential to reduce GHG emissions in the country's natural gas sector, as well as to limit the flaring of gas associated with oil extraction. The economic value of the saved gas justifies the identified energy-efficiency improvements, which would also enhance energy security for Russia and importing countries. It will thus reinforce Russia's role as a reliable supplier of natural gas in the coming decades. However, structural and regulatory reform is needed to ensure the efficiency of Russia's gas sector, as well as to enable an effective implementation of Kyoto Protocol mechanisms.

OVERVIEW OF GHG EMISSIONS:
RUSSIA'S NATURAL GAS SECTOR AND GAS FLARING

In 2004 Russia emitted an estimated 298 million tonnes of CO_2 equivalent ($MtCO_2e$) of GHG from its natural gas transmission and distribution systems, and through gas flaring, about 15% of the country's total GHG emissions (see Table ES-1).[2] In 2004, just under 70 bcm, equivalent to just over one third of Russian exports, either leaked in the form of methane (CH_4) from various components along Russian transmission and distribution pipelines in normal operations, was used as fuel gas in the transmission process, or was flared by oil companies. Although over half of this volume was used by compressors along the gas transmission system, significant efficiency improvements are still feasible in this area, in light of comparable systems in other countries.

The transmission sector accounted for about 60% of total GHG emissions while the gas distribution network accounted for over a quarter. Gas flaring emissions by oil companies accounted for 14% of total according to official data. CH_4 emissions accounted for about 60% of total GHG emissions and were due to leaks from pipelines and compressors during normal operations, maintenance, repairs, and accidents.

Table ES-1   Estimated GHG emissions: Russia's natural gas sector and gas flaring in 2004

	Gas combustion and leaks, bcm	GHG emissions, $MtCO_2e$	Structure of GHG emissions
CH_4 leaks from transmission pipelines and compressors	6.2	93	31%
CH_4 leaks from the distribution system	5.3	80	27%
CO_2 emissions from gas combustion at compressors*	41	82	28%
CO_2 emissions from flaring of associated gas**	15	43	15%
Total	**67.5**	**298**	**100%**

* Gas consumption by the transmission system which can be reduced through more efficient compressors
** Emission factors differ due to the higher share of "heavier" gases in associated gas and incomplete combustion

---

2. Without official sectoral inventories and the lack of information due to the insufficiency of meters, these estimates should be considered as orders of magnitude.

RUSSIAN CLIMATE POLICY

Russia could play a determining role in Kyoto markets

Russia's emissions targets in the Kyoto Protocol leaves it with a surplus of transferable emission quotas of 330 to 800 $MtCO_2e$, a potentially important contribution to the Annex I countries' compliance with their Kyoto commitments, in light of their current emission trends. The Russian government has expressed its interest in linking all traded quotas to measurable emission reductions, a way to secure the "environmental integrity" of its transactions and attract demand. This is feasible under the protocol's project-based mechanism (so-called Joint Implementation), yet Russia must move swiftly to become eligible for such a mechanism. It will otherwise need to submit each GHG reduction project to international scrutiny – the so-called Track 2 of Joint Implementation – arguably a more burdensome and transaction cost-ridden procedure. Russia has also proposed to sell existing surplus quotas and reinvest revenues in GHG reduction projects – via a so-called Green Investment Scheme.

Slow implementation of Russia's Action Plan on Kyoto

There are visible signs of the Russian government's progress in developing its climate policy. The ongoing debate seeks to identify the set of instruments that could enhance the synergy between climate policy and long-standing, but so far largely unsuccessful, energy-efficiency policies. An Action Plan sets out very ambitious timelines to make the country eligible for the Kyoto Protocol emissions trading mechanisms, an objective within the reach of Russia's technical and financial resources. After some delay, new deadlines were set in March 2006 and the government reconfirmed its commitment to completing the process before 2007.

Need for clear signals and rules for investors

Russian industry and environmental organisations support the development of Joint Implementation (JI) in Russia, as it could improve investment returns on energy-efficiency projects through the revenue stream attached to GHG emission reductions. Russian companies are developing their investment proposals and foreign investors are expressing their interest, including in the natural gas sector. However, investors are waiting for clear signals from the government on its climate policy and on various ministries' responsibilities in its implementation.

Project-based mechanisms to "boost" investment in less attractive sectors

Given the transaction costs related to JI Track 2, large scale projects may be taken up first. This would nevertheless provide a first step in establishing a climate policy framework in Russia. JI could be an option for investment projects in capital-intensive sectors such as gas transmission, harbouring significant volumes of GHG emission reductions. Going through a project-by-project approach avoids having to wait for more comprehensive sector-wide GHG inventories.

Russian authorities have shown a preference for the use of Kyoto mechanisms in sectors less attractive to investors, such as the residential sector (district heating) as well as the coal sector (with a focus on coalbed methane), where GHG reductions can still be significant. The gas distribution sector could be prioritised in such an approach, as it needs technical and financial capacity to perform much needed network upgrades.

Russia's Emissions Trading Scheme not expected before 2010	In the longer run, the establishment of a national emissions trading scheme (ETS) in Russia would allow a larger number of national emitters to access the market for GHG reductions and related finance. A domestic ETS linked to other systems could facilitate international transactions for Russian investors. However, such a system will need energy sector reform to perform its role as a price signal for efficient GHG reduction investments. Experience in IEA countries has also revealed the major administrative effort needed to implement emissions trading. Russian experts and officials, and foreign observers do not deem implementation of such a system to be feasible before 2010.
Timely establishment of a GHG emission inventory in Russia	Methane (CH_4) emissions represent a considerable share of the GHG emission reduction potential in Russia's gas transmission and distribution systems. Reliable CH_4 emission estimates for the natural gas sector are essential to validate conventional wisdom, and to highlight opportunities for investments via the Kyoto flexibility mechanisms. The establishment of a sectoral GHG emission inventory will provide useful insights to determine baseline scenarios, from which CH_4 emission reductions can be assessed.

RUSSIAN GAS TRANSMISSION SECTOR

Low efficiency of compressors and ageing transmission system	Russia's rather inefficient gas transmission system is a large emitter of GHG. In comparison with foreign gas systems, its high energy intensity is due mainly to the large number of low efficiency compressor units along Gazprom's transmission system and to the ageing of its facilities.
Lack of refurbishment in the past reduced the transmission system's capacity	In 2004, almost 700 bcm of natural gas flowed through Russia's high-pressure transmission system, including imports and transit from Central Asia. Due to under-investment in maintenance and repair during the 1990s until 2002, investment in refurbishment is long overdue in Gazprom's transmission system. In 2002, Gazprom had to reduce the throughput of the system to 60 bcm less than its rated operational capacity. The lack of spare transmission capacity has limited third party access and the development of domestic competition in the upstream gas sector, while Gazprom seems more keen on investments in export pipelines than on refurbishing domestic transmission.
Investments could lead to annual gas savings of up to 10 bcm	Gazprom estimates that gas consumption and losses in its transmission system can be reduced by 10% to 20%, up to 10 bcm per year. Such improvements could bring about reductions in GHG in the order of 50 $MtCO_2e$ per year by 2012. These savings are available at low and medium upfront cost; and are economic thanks to the corresponding increase in gas sales on the domestic market, let alone the international market. The long-standing partnerships between Gazprom, EU, Canadian and Japanese gas companies could foster the implementation of such savings, which should be tapped as domestic prices increase, and Gazprom's three key producing fields decline.

RUSSIAN GAS DISTRIBUTION SECTOR

Lack of investment for upgrading and maintenance

In contrast to the many studies and international projects already undertaken on Gazprom's transmission pipeline system, very little attention has been given to GHG emissions in Russia's gas distribution network. With 575 000 km of distribution pipelines, it ranks as the world's second largest system after the United States, distributing over 380 bcm of natural gas to the domestic market in 2004. About 40% or 170 bcm is assumed to be supplied by medium and low-pressure distribution networks.

Low end-use tariffs and lack of meters hamper incentives for energy savings and upgrades. Ageing and lack of maintenance make things worse. In Russia, only one quarter of the pipelines at the end of their operational life (40 years) are monitored annually.

The inherent problems of lack of metering and the "imbalance"

The lack of meters and monitoring devices raise a special problem in Russia which they describe as "imbalance", *i.e.* the difference between the volumes of natural gas supplied and those recorded as consumed. This difference may result from fugitive emissions during normal operations, accidents or theft. Some experts argue that the share of theft or "commercial losses" could account for as much as 70% of the "imbalance". Gazprom experts point to the lack of meters or the use of old faulty meters as the key factor of this "imbalance". The various reasons behind this "imbalance" could make precise audits of losses politically difficult.

The installation of meters is an urgent necessity for gas distribution companies to tap their energy-saving potential. This could also foster a more efficient use of energy at end-use level.

Huge, but dispersed GHG emission reductions potential in the distribution sector

Recent studies and pilot projects in certain Russian regions have shown that potential reductions of GHG emission from ageing and badly maintained distribution networks may be as large as in the transmission sector. About 3% of the total gas distributed by medium and low-pressure pipelines is estimated to be leaked into the atmosphere – about 80 $MtCO_2e$. Projects to reduce such leaks are necessarily small and dispersed over hundreds of municipal systems. Transaction and monitoring costs may render these projects unattractive. The limited financial and technical resources of gas distribution organisations are other barriers to overcome.

Kyoto flexibility mechanisms could help overcome investment barriers

The analysis of CH_4 emission reduction potential of Gazprom shows a significant amount of low cost options. These options could be implemented during maintenance and repair programmes and become an integral part of common practice in the Russian gas distribution sector, if financial and technical capacities were available. Kyoto flexibility mechanisms could be useful and timely in overcoming these barriers and attracting essential investments in this sector. Projects could be grouped to bring economies of scale and lower transaction costs.

GAS FLARING REDUCTION

In 2005, 15 bcm of gas were officially flared

In 2005, Russia officially reported that 15 bcm of associated gas were flared, an underestimate according to international and Russian experts. The problem of gas flaring and the lack of transparency is not specific to Russia and is increasingly highlighted by governments and industry, and in particular by the World Bank.

True third party access: a key to reducing gas flaring

Apart from purely economic obstacles to the efficient use of associated gas, a key issue in Russia is the current structure of the Russian gas sector. Russian oil companies are increasingly interested in raising their production of oil and therefore associated gas, but are hampered by Gazprom's monopoly of the gas pipeline network and lack of commercially viable access to gas processing facilities.

A new tool to estimate gas flaring volumes to build on the World Bank's initiative

The IEA, together with the National Oceanic and Atmospheric Administration of the United States (NOAA), calibrated various satellite images of flares in West Siberia against a known sample flare. Preliminary estimates based on this method indicate flaring of 60 bcm, over 4 times the official figure. More benchmark data points would be necessary to obtain more accurate results. However, finding this data has proven difficult. The IEA recommends more transparent information on gas flaring volumes in Russia and around the world to allow such analysis.

Carbon finance can help enhance the economics of projects to use associated gas

Our rough assessment of the "typical" options to enhance the use of associated gas reveals that the economics of various options are limited by many factors including low domestic gas prices, the distance between production and consumption points, the limited gas needs of oil companies, the costs of necessary infrastructure and low associated gas flow rates. For gas re-injection projects, as well as for emerging gas-to-liquids options, carbon finance *i.e.* additional financing through Kyoto mechanisms, could provide a guaranteed revenue stream and enhance project economics. In projects to move associated gas to markets through pipelines, the impact of a carbon revenue component on investment decisions may not be enough to overcome the uncertainty of long-term reliable access to Gazprom's pipelines.

The synergy between carbon finance and structural and market reform

Oil companies, currently benefitting from high oil prices, should not find it financially difficult to invest in projects to use associated gas, all the more so as such projects can provide them with another revenue stream without carbon finance. For this reason, the Russian government has not given priority to such investment projects in its climate policy. However, the current structural and market barriers described above may limit oil companies' interest in gas flaring reduction projects in Russia, especially given other investment options.

The barrier approach to demonstrate "additionality"

This should be taken into account when determining the *additionality* of gas flaring reduction projects, *i.e.* the extra environmental benefit on top of what would otherwise be the case without the Kyoto Protocol. Some gas utilisation projects may be considered additional even if they aim to comply with the mandatory limits established in the field licenses – these limits are essentially unreachable in current conditions.

OVERALL ASSESSMENT

Each sector of Russia's gas industry could contribute to GHG emission reductions in a cost-effective way. Figure ES-1 summarises our results, showing on the left side estimated CH_4 leaks along Russian transmission and distribution networks and on the right side estimated CO_2 emissions due to combustion of natural gas at compressor stations along the transmission network and flaring of associated gas by oil companies. Equivalent CO_2 emissions show the large contribution of gas leaks to this total.

Figure ES-1 also indicates our estimates of available reductions in emissions and combustion of natural gas in the various sectors. Over 60% of GHG emissions can be reduced along transmission and distribution pipeline systems. The IEA estimates that more than half of the potential reductions of CH_4 leaks could be found in the distribution network. However, this sector encompasses the most uncertainty. In terms of natural gas savings, the volumes are much less significant than the volumes which can potentially be reduced at compressor stations along the transmission system or flared by oil companies.

Flaring activity is far from transparent in Russia and hinders a definitive estimate. The IEA assumes that the totality of currently flared gas could be used economically, albeit not under current third party access conditions to Gazprom transmission infrastructure. The current situation is hardly efficient in light of the negative effects on overall gas supply volumes and the global environment.

Figure ES 1   Estimated GHG emissions in 2004 and potential for reductions in Russia's natural gas sector and gas flaring

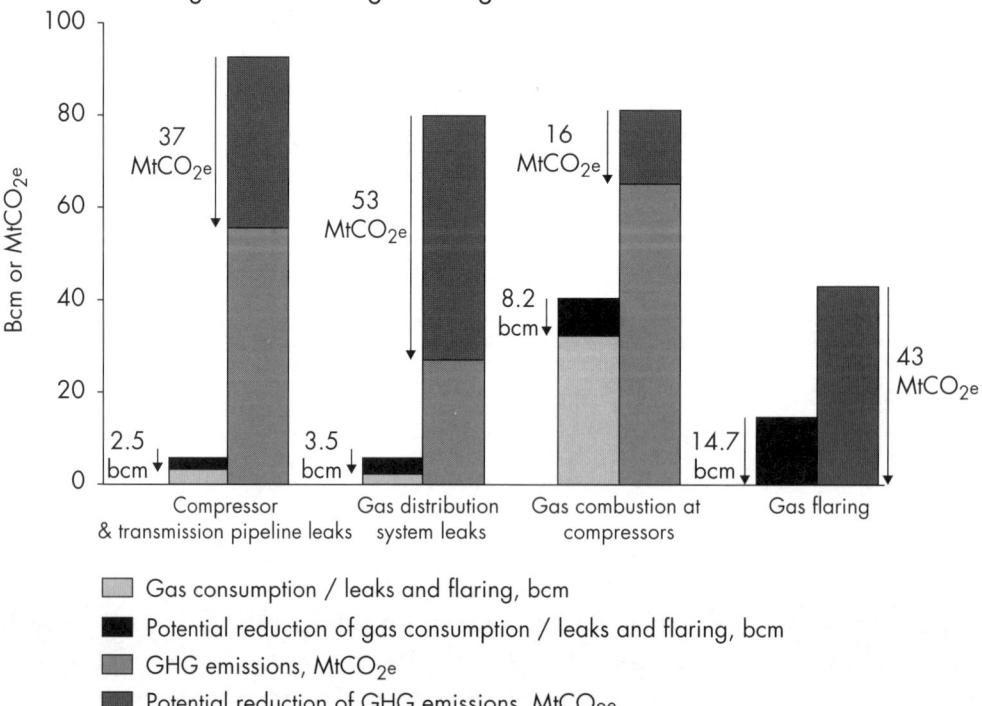

Climate policy needs to be backed by sectoral reforms

The Russian government is currently seeking to establish an efficient GHG reduction policy system. Success of emission reductions in Russia's gas sector, with the assistance of the Kyoto Protocol mechanisms, will largely depend on the implementation of gas sector reforms. Independent of climate policy, these structural and regulatory reforms will fundamentally determine: i) the priority or strategic choices of main actors such as Gazprom, ii) the efficiency of gas distribution companies and iii) the investment decisions of oil companies with respect to associated gas.

Access to Russia's low-cost GHG emission reductions potential, namely in the gas sector, is of great interest to OECD companies and governments, who must achieve their international GHG emission reduction commitments and are keen to demonstrate environmental benefits.

Carbon finance is not a silver bullet to overcome sectoral and market barriers

Throughout this book the IEA emphasises that without reforms in the gas sector the effectiveness of the Kyoto mechanisms to enhance GHG emission reductions in Russia could be severely limited – as illustrated by the lack of third party access to transmission which encourages gas flaring. The price of CO_2 alone cannot be effective without regulatory reform in this area.

I. RUSSIA'S NATURAL GAS CHAIN : OVERVIEW AND OUTLOOK

Russia's gas resources are huge. It has 47 trillion cubic meters of proven natural gas reserves – 26% of the world total (Cedigaz, 2004). In 2004, Gazprom held the licences to fields accounting for 60% of these reserves and controlled the infrastructure essential to the rest. Three-quarters of Russian gas reserves – and a similar share of current production – are in West Siberia, mostly in the Nadym-Pur-Taz region. European Russia (including the Barents Sea shelf) holds 16% of reserves and East Siberia and the Far East together have the rest.

Russian gas production fell sharply in the 1990s in response to the decline in domestic demand following the break-up of the Soviet Union, from a peak of 643 bcm in 1991 to a trough of 571 bcm in 1997. Production has since recovered, largely for export. Production reached 641 bcm in 2005, of which Gazprom produced 547 bcm predominantly from the three super-giant fields in Nadym-Pur-Taz now in decline: Medvezhye, Yamburg and Urengoy. Output from a fourth super-giant field, Zapolyarnoye, which started producing in 2001, reached its peak of about 100 bcm in 2005. Beginning in 2008, the cumulative decline in production at Gazprom's three major producing fields will be greater than the current peak production at Zapolyarnoye – the last relatively cheap gas in Gazprom's portfolio. With the era of relatively "cheap" gas and inherited Soviet gas infrastructure over, Gazprom is at a crossroads of major decisions and choices:

■ Developing Russia's extensive smaller gas fields would require a robust, competitive gas sector. This is not the case.

■ For the larger Yamal fields including its pipeline infrastructure, costs are much higher than its currently producing mega fields and plans have been delayed for 10-12 years.

■ Developing the Shtokman field as Gazprom's first LNG (liquefied natural gas) project will require huge investments and deep water development technologies unavailable inside Gazprom.

■ Developing East Siberian natural gas fields and the necessary transportation infrastructure will be very expensive.

■ Providing better pipeline access to independent gas producers and oil companies producing associated gas as well as natural gas fields of their own is an option Gazprom has not chosen to date.

■ Increasing imports from Central Asian countries is becoming increasingly problematic both because the linked pipeline networks built during the Soviet era are in dire need of refurbishment and expansion and also because Central Asian producers, aware of the above points, are therefore in a stronger negotiating position.

Russia needs to address these important points in the coming years to maintain its position as a secure and reliable supplier of natural gas both to its domestic and export markets – especially as Russia considers opening new pipeline and LNG markets to the East and to North America. In short, is there enough investment being made in Russia in natural gas exploration, development and infrastructure?

If Russia could tap its energy-efficiency potential as its economy continues to develop and grow, the problems above would become more manageable. The Russian Energy Strategy reflects a keen understanding by Russian energy policy makers that Russia's energy-efficiency potential is huge and needs to be tapped, estimating that Russia could reduce consumption of energy per unit of GDP by between 39-47% from 2000 levels. Cost-reflective pricing of energy will be needed to create incentives to energy efficiency.

Moreover, Russia's implementation of the Kyoto Protocol can help in realising the synergies between limiting GHG emissions and more efficient consumption of Russia's natural gas. This study examines the potential of reducing GHG emissions in Russia's natural gas sector and limiting the flaring of associated gas by Russian oil companies. This book will show that through the use of Kyoto flexibility mechanisms, Russia could substantially improve the energy efficiency of its natural gas sector thereby enhancing energy security and its ability to remain a reliable supplier of natural gas in the coming decades.

NATURAL GAS BALANCE: TRENDS, EXPORTS AND OUTLOOK

Gas supply

Over a decade ago, Russia's gas sector looked very different. Russian gas production was declining steadily from a level of 643 bcm in 1991 to 584 bcm in 2000. Gazprom production had declined from 595 bcm and 523 bcm, with independent gas producers making up the rest. During this time, Russian energy policy makers released various versions of a new energy strategy – each reflecting a lower outlook for future Russian gas production. The 1995 Energy Strategy estimated a maximum requirement of 860 bcm and a minimum of 740 bcm in 2010, yet by 1999 the new Energy Strategy estimated a lower figure of 700 bcm of natural gas for 2010. The final Energy Strategy approved by the government in August 2003 set Russian gas production needs at between 635-665 bcm in 2010 increasing to 680-730 bcm in 2020. However, the production requirements of 610-615 bcm for 2005 were significantly underestimated – as actual production during this year reached 647 bcm.

Were energy policy makers drafting the Energy Strategy influenced by the decline in natural gas production over the 1990s? Were they influenced by analysis showing the huge potential for energy efficiency? Were they influenced by the thinking at that time of a need to refocus Russia's total primary energy supply (TPES) away from what was considered an over-dependence on natural gas and a need to rebalance with an increased use of coal? This latter concept of gas-to-coal switching was introduced by Gazprom in an effort to reduce domestic consumption of gas – both because of non-payment of low domestic gas prices, and also because this would allow more gas to reach hard currency export markets.

In any case, Russian domestic gas consumption continued to increase, paying no heed to the concept of gas-to-coal switching with close to 70% of thermal electricity generation in 2004 continuing to be fuelled by gas. What had changed since the drafting of the Energy Strategy? Independent gas producers appeared to take up a growing share of the domestic gas market. They were willing to do this as their only other market was selling gas to buyers at unregulated gas prices when Gazprom did not have sufficient gas. As domestic prices increased, this market became attractive to these small independent gas producers, with no overheads and low development costs. As demand has increased with economic growth, this market is becoming more interesting.

The key constraint to independent gas producers in Russia was and continues to be access to the gas pipeline system controlled by Gazprom. They hope that as Gazprom sees the benefit of independents meeting growing domestic demand, more reliable access should ensue. This could accelerate if a draft regulation submitted by the Federal Anti-Monopoly Service in early 2006 is designed to promote third party access to Russia's natural gas transmission system. It proposes auctions for access to gas pipelines, new terms for gas transportation contracts and better access to information on spare pipeline capacity.

Gazprom

In June 2003, OAO Gazprom, which accounts for 90% of Russian gas production, announced a completely new strategy for the company "From Stabilisation to Growth" at its Board of Directors meeting. The Strategy projects Gazprom's production levels to increase from 547 bcm in 2005 to 560 bcm in 2010, 590 bcm in 2020 and 630 bcm in 2030. Gazprom's strategy, consistent with the Russian government's overall energy strategy, anticipated the share of independent gas production increasing from 15% in 2005 to 20% in 2020.

However, to achieve these production targets, Gazprom will have to increase its annual reserve replacement in the order of 700 bcm/y to 2015 and 750-800 bcm/y for 2016-30.[3] This is 36% more than the 2002 reserve replacement level, the last time in almost a decade when reserve replacement was anywhere near production. Gazprom's reserve replacement dropped to 79% in 2003, 69% in 2004, and just over 100% in 2005 (see Figure 1).[4]

Much of Gazprom's current production is from Cenomanian deposits, with production costs estimated at about USD 10/thousand m³. In 2001, Gazprom commissioned the Zapolyarnoye field (3.3 trillion m³) which reached full capacity of 100 bcm/year production in 2005. It is considered by Gazprom as its last relatively cheap gas reserve. Future fields will be in more difficult-to-develop regions or in deeper

---

3. From a Gazprom meeting in Sochi in April 2004 where Gazprom's development strategy was discussed.
4. It is not clear, however, if this was due to Gazprom's acquisitions of independent company reserves, *i.e.* inorganic growth – or reserve accounting related to its swap with Royal Dutch/Shell where Gazprom entered into the Sakhalin II project with a 25%+1 share in return for a 50% stake in the Zapolyarnoye field.

Figure 1   Gazprom reserve replacement

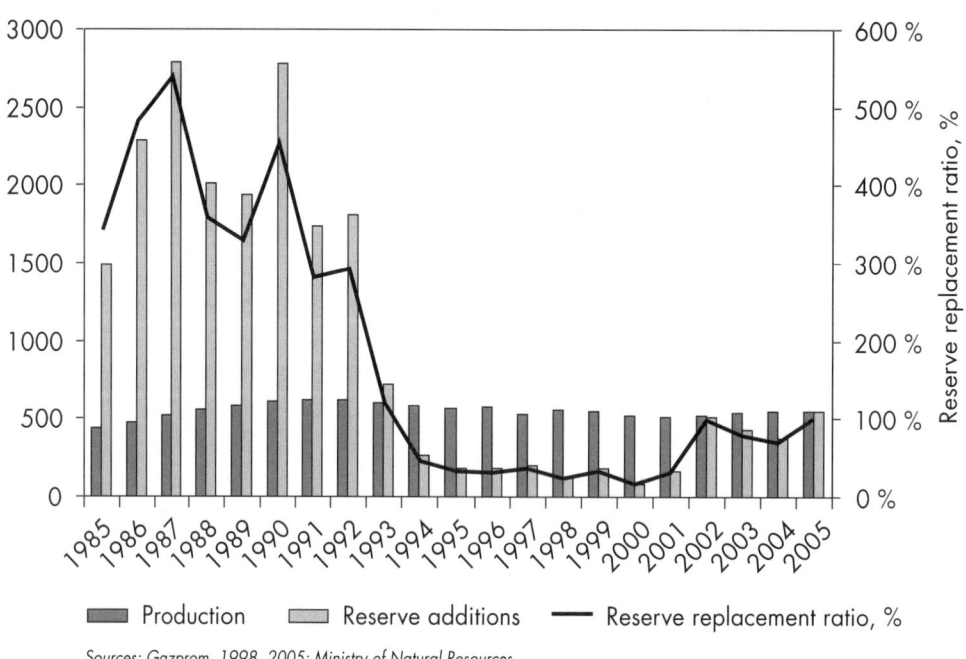

Sources: Gazprom, 1998, 2005; Ministry of Natural Resources.

geological levels, more expensive than Cenomanian plays. The Russian Energy Strategy presents cost estimates for development of the Yamal fields in the order of USD 30/thousand m³ excluding transportation infrastructure. Over the early 2000s, Gazprom pointed to preparatory work that had already been carried out in the Yamal peninsula – Bovanenkovskoye (4 trillion m³), Kharasavsky (1.3 trillion m³) – as well as starting-up the development of the Kamennominskoye shelf and North-Kamennominskoye (0.8 trillion m³). However, it appears that the development of Yamal by Gazprom changed dramatically in early April 2006, when a Deputy Chairman of Gazprom stated that the Yamal fields would not be developed before 2016-18.[5]

Judging by this recent statement, it appears that those within Gazprom supporting a more active role in LNG markets may have taken the lead – at least as far as the longstanding discussion on whether to develop Yamal fields before Gazprom's other huge field, Shtokman (3.7 trillion m³) is concerned. In mid-September 2005, Gazprom short-listed five international majors as its potential partners for the Shtokman development in the Barents Sea. The final partners were to be announced by end-April 2006, however, this was put off and timing now is unclear. This project could be implemented under a Production Sharing Agreement (PSA), with the aim of starting deliveries from an LNG plant to North American markets after 2010, and has been accorded "strategic" status meaning that it will be 51% controlled by Gazprom.

Gazprom has recently taken an interest in Russia's other potential gas regions in East Siberia, the Far East and Sakhalin. In July 2005 Gazprom and Shell signed a

---

5. RIATEC, 2006.

Memorandum of Understanding whereby Gazprom would acquire up to 25% plus one share in the Sakhalin II venture, and Shell would acquire a 50% interest in the Zapolyarnoye Neocomian field in West Siberia.[6] Gazprom, as supervisor of the development in East Siberia and the Far East, has asserted its intention to participate in all natural gas developments in the region to ensure its control of export routes and volumes.[7] Gazprom's intentions seem to indicate that Kovykta and Sakhalin gas resources belong to international consortia in which Gazprom has only a share – presumably these consortia will determine the markets for their gas.

The Russian government is intent on developing this scarcely populated vast region. Russia has a long standing declaration of intent to co-operate with China given the East Siberian oil and gas resources and China's interest in importing increasing volumes from its neighbour. This was discussed at the highest levels in spring 2006 when inter-governmental framework agreements were signed by President Putin of Russia and President Hu Jintao of China. President Putin stated that Russia could potentially supply an annual total of 60-80 bcm of gas to China using eastern and western routes which would each supply 30-40 bcm. Gazprom stated that the planned USD 10 billion 3 000 km Altai pipeline system (the western route) would pump the first Russian gas to China as early as 2011. Gazprom's President also said that the Kovykta field in the Irkutsk region of East Siberia could be a possible export source – but that gas from Sakhalin or West Siberia was still being considered. These political statements made in spring 2006 are very ambitious, especially with Gazprom's increasing assertion of control in this region over recent years, which has done little to spur on development by private investors in the region.[8] This begs the question as to the intent behind these statements. Clearly there is a political will – but is the timing of these statements more a reflection of concerns being raised in Europe on Russian export markets?

Gazprom annual investments have been in the order of USD 7 to 8 billion since 2003. In 2005, Gazprom's management board approved a more than 40% increase in its investment programme to USD 10.8 billion, much of the increase being directed to the North European Gas Pipeline project. This corresponds to the IEA's estimate in its *World Energy Investment Outlook* (IEA, 2003) of USD 11 billion per year required to bring on new sources of gas and to upgrade and maintain gas infrastructure. The IEA is concerned about the priority Gazprom seems to be placing on foreign acquisitions and export infrastructure as opposed to its domestic network and upstream.

Central Asian countries

Gazprom's transmission system, built during the Soviet era, focussed not only on its key producing fields in West Siberia, but also extended to fields in Soviet Central Asia. The United Gas Supply System of the Soviet Union was built on the basis of the natural gas reserves of West Siberia and Turkmenistan, Uzbekistan and Kazakhstan,

---

6. The transaction was expected to be finalised in 2006. However, shortly after the Memorandum was signed, Shell announced that the costs of the second phase of the Sakhalin II project had risen from USD 12 to USD 20 billion. This gave Gazprom grounds to renegotiate the terms of the asset swap and discussions are on-going.
7. See Gazprom's Web site for more details on its regional programme for East Siberia and the Far East at www.gazprom.ru/docs/topics/2677.shtml and www.riatec.ru/shownews.php?id=23705.
8. These intentions seem to set aside that Kovykta and Sakhalin gas resources belong to international consortia on which Gazprom has only a share. Presumably these consortia will determine the markets for their gas.

Map 1 Major natural gas producing and prospective regions and pipelines

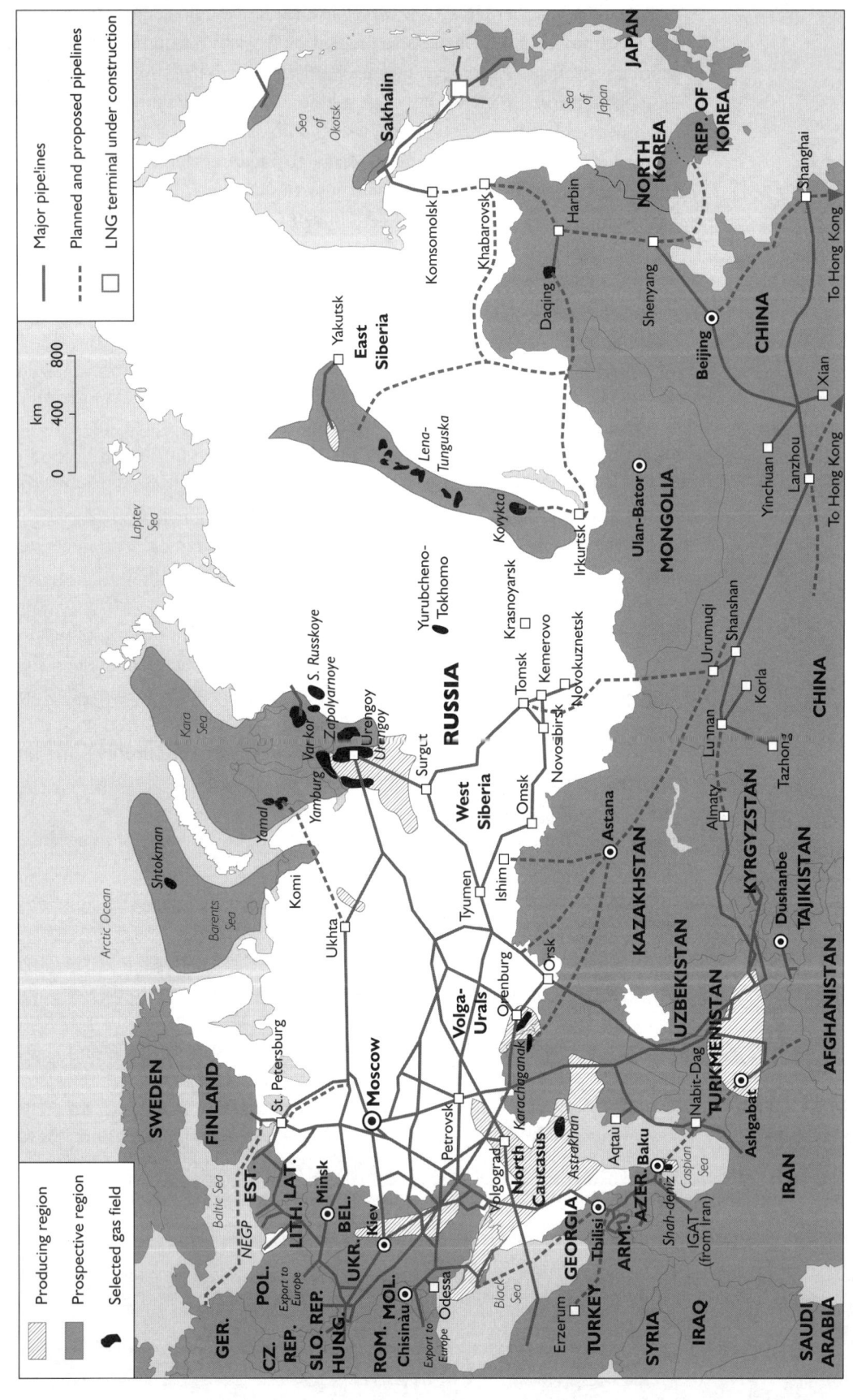

Source : IEA.

then part of the Soviet Union (see Map 1). With the depletion of its key Siberian fields and the prospect of more expensive production from its remaining assets, Gazprom's upstream strategy was to draw on the cheaper gas in Central Asia by using existing pipeline infrastructure.

In April 2003, Gazprom signed a long-term framework agreement with Turkmenistan for the purchase of 5-6 bcm in 2004, rising to 6-7 bcm in 2005, 10 bcm in 2006, 60-70 bcm in 2007 and 70-80 bcm over 2009-28.[9] The sharp increase in 2007 coincided with the expiry of the existing 36 bcm/y supply agreement between Turkmenistan and Ukraine at the end of 2006. However, over 2005 and into 2006, this strategy to import increasing volumes of gas from Turkmenistan ran into pricing difficulties. At present, the level of Russian imports from Turkmenistan is insignificant, accounting for less than 1% of total production. However, Gazprom's plan for Turkmen volumes to increase dramatically must make Gazprom and the Kremlin uneasy about the terms and availability of supply – even if Europe appears quite relaxed.

The IEA has long questioned the advisability of relying on contracts between Russia and Turkmenistan to meet future increases in European gas demand. These agreements and alliances formed with Central Asian countries allowed Gazprom to delay any competition in Russia while effectively removing Central Asian gas as a potential competitor in European markets. Gazprom has delayed development of capital-intensive and increasingly expensive reserves in Yamal and the Barents Sea. The policy of relying on Central Asian volumes has postponed addressing the more fundamental problem of compensating for the decline of its major fields, and the need for reform of the Russian gas sector.

Pipeline infrastructure: Russia and Central Asian countries

The Central Asia Centre (CAC) pipeline network made up of 5 different lines was designed and built over 1966-87 with an overall capacity of about 90 bcm/y. Four lines of the system originate in Turkmenistan and pass through Uzbekistan with the fifth branch through Kazakhstan. The lack of maintenance and investment over time has almost halved the operational capacity of the system to about 50 bcm/y. If Russia intends to increase Turkmen exports to 80 bcm/y, not to mention the expected increase in exports from Kazakhstan (15 bcm) and Uzbekistan (10 bcm), major refurbishment and expansion of the CAC system will be necessary. Gazprom has made its financial support of an expansion of the CAC contingent on the release of an independent audit of Turkmenistan's gas reserves. Gazprom is particularly interested in expanding the eastern branch of the CAC which runs through Uzbekistan, in order to pump additional gas that it is planning to produce there under several production and supply agreements. Upgrading the Kazakh part of the system is estimated by KazTransGas to cost USD 2 billion. Uzbekistan estimates investments needed for refurbishment of its lines at about the same price. The Turkmenistan government estimates the cost of refurbishment at less than USD 1 billion. All of these numbers are likely to be low.

---

9.   Since tensions in January 2006 with Ukraine, Turkmenistan is supplying 30 bcm of natural gas to Russia in 2006 at a price of USD 65/thousand m^3. Russia is interested in increasing this volume to 50 bcm, but all depends on the on-going price negotiations.

The Russian part of the transmission system is controlled by Gazprom, and although in a better state of repair, it also suffers from reduced capacity throughput along various chokepoints. The system was built mainly between 1975 and 1990, when the massive increase in gas production from West Siberia occurred. Most of the export pipelines are more recent and have better technical parameters than those of the domestic transmission system. In 2002, experts at VNIIGAZ estimated that the operational capacity of Gazprom's transmission system was 60 bcm less than its designed rated capacity as many pipes could no longer withstand design pressures (Leontiev and Stureiko, 2003; Pravosudov, 2004a).

As discussed in Chapter 3, in 2002-06, Gazprom undertook its third comprehensive 5-year programme to refurbish its trunk gas pipeline system including its compressor stations and gas storage facilities. Investments undertaken in this latest programme have been much closer to target levels than in the past. Throughput volumes in 2004 were up 6% in comparison to 2001 levels. The overall goal was an increase in rated throughput capacity by 35 bcm/y and a decrease in energy input fuel needs for the transmission system of 5 bcm/y. This is particularly significant as in the past Gazprom used the lack of spare capacity as justification to deny third party access to its transmission system. This continues to limit the development of gas production by independent gas producers and Russian oil companies.

Independent gas producers and oil companies

Russian oil companies and independent gas producers hold just under a third of Russia's gas reserves and are capable of making a growing contribution to Russian gas production in the coming decades as partially reflected in Ministry of Industry and Energy projections. They already account for an estimated 13%, all of it sold to domestic customers at lower domestic prices now averaging USD 40/thousand m^3 compared to the European price of USD 260/thousand m^3. Several companies are seeking to boost production, much of it associated with oil. Were it not for the reserves in Turkmenistan, and to a lesser extent in Uzbekistan and Kazakhstan, there would surely have been a more receptive ear from both Gazprom and the Russian government to the growing Russian independent gas producer and oil company lobby over the past few years for more transparent and reliable access to Gazprom pipelines.

Company projections imply that total non-Gazprom output could reach 260 to 290 bcm by 2015 – about 40% of total Russian gas production (see Table 1). However, the Russian Energy Strategy projects only half this amount from independents, accounting for only 20% of total production in 2020.

Third party access has improved since 1998 when only 6 independent organisations (28.2 bcm) gained access. By 2000, 20 independent organisations with volumes of 106.2 bcm were allowed access. This number increased to 33 organisations by 2004, although the volume of independent throughput dropped to 99.9 bcm of natural gas. This increased access over recent years reflects the activities of third parties importing or transiting natural gas from Turkmenistan to Ukraine as well as long-term contracts signed by Gazprom for associated gas from independents and oil companies such as Lukoil (with whom it signed a long-term contract in 2003). In mid-2005, Gazprom signed a long-term agreement with Novatek, the largest of the independent gas producers. Although this eases pipeline access, it does not foster competition in the upstream Russian gas sector or beyond.

Table 1   Russian gas production by oil companies and independents in 2003-05

	2003	2004	2005	Company expectations 2010-2015
Oil companies, bcm	*40.5*	*44.9*	*49.0*	*215*
Surgutneftegaz	13.9	14.3	14.4	25
TNK-BP	6.8	8.0	8.7	20-40
Rosneft	7.1	9.4	13.0	50
Yukos	3.4	3.4	2.0	50
Lukoil	4.7	5.0	5.8	50
Sibneft	2.0	2.0	2.0	-
Other	2.6	2.8	3.1	-
*Independents, bcm**	*36.9*	*45.8*	*45.0*	*75*
Novatek	21.0	27.8	25.4	52
Nortgaz	5.0	3.7	3.2	11
PSAs (including Sakhalin)	0.2	0.3	0.5	12
Other	10.7	14.0	15.9	-
Total, bcm	**77.4**	**90.7**	**94.0**	**260-290**
Russian Energy Strategy				115-135

* Expectations of independents are all for 2010
Sources: IEA estimates; Oil company reports; Energy Strategy, 2003.

Flaring of associated gas: lack of third party access

The prospects for independent gas production depend on transparent and reliable access to Gazprom's gas-processing capacity and transmission system. Large volumes of associated gas produced by oil companies are still being flared. In some cases the use of associated gas is uneconomic due simply to the long distance between production and consumption points, or to geophysical difficulties related to re-injecting the associated gas into the field without diminishing oil recovery. In Russia, however, the issue is more often related to Gazprom's dominant position which enables it to deny access to its transmission system. Claiming capacity constraints, Gazprom refuses to buy associated gas from oil companies or independents. The terms of access to Gazprom-controlled gas processing plants can also pose possible hurdles for those without treatment plants, thus rendering projects based on associated gas uneconomic.

As Russian domestic gas prices increase, fields close to main gas transmission pipelines and gas processing infrastructure will have a greater interest in marketing their gas. This is a key focus of our study and depends entirely on Gazprom's willingness to provide commercially attractive access to its transportation system. This is discussed in more detail in Chapter 5. We also assess the potential of national climate policy as a force for implementation of third party access to Russia's gas network.

Gas supply security: for Russia and its export markets

The lack of competition in Russia's upstream gas sector is increasingly disconcerting given the tension in gas supplies to European customers already apparent during the extraordinarily cold weather in early 2006. This reflects the technical limits of Russian gas production and its transport capacity. With Gazprom's major fields in decline, and its unwillingness to undertake or authorise other domestic options, Russia relies on Central Asian gas to meet the growth in its contracts with Europe. But is there sufficient investment in Central Asian gas? Current IEA projections suggest that Gazprom could face a gradually increasing supply shortfall against its existing contracts beginning in the next few years if timely investment in new fields is not made.[10] With a lack of convincing information from Gazprom to the contrary, the IEA projects an average 20 bcm/year natural decline in Gazprom's production reflecting historic decline rates in its big 3 fields. At this rate, by 2015, almost 200 bcm will need to be produced from new Gazprom fields, if it is to maintain production at current levels – let alone meet its new strategy goals of increasing production to 560 bcm/y in 2010 and 590 bcm/y in 2020. Figure 2 also assumes no growth in imports from Central Asia beyond current levels. This estimate shows the apparent lack of investment in Russia's upstream by Gazprom or in Central Asia's upstream and mid-stream.

The need for Gazprom to invest in new more expensive fields is all the greater if Russian imports from Central Asia are not forthcoming for either political or supply reasons. Little information on current investments in this pipeline infrastructure is available publicly, although the information available from oil companies, paints a bleak picture of investment and/or stability in the Central Asian gas sector. The only promising information is that provided by Lukoil which is planning to produce and export about 10-12 bcm of gas from Uzbekistan in coming years from its Kandym project.

The IEA's outlook of Russian natural gas production incorporates projections of the Russian Energy Strategy related to independent gas producers. However, as shown

Figure 2   Russian gas supply outlook

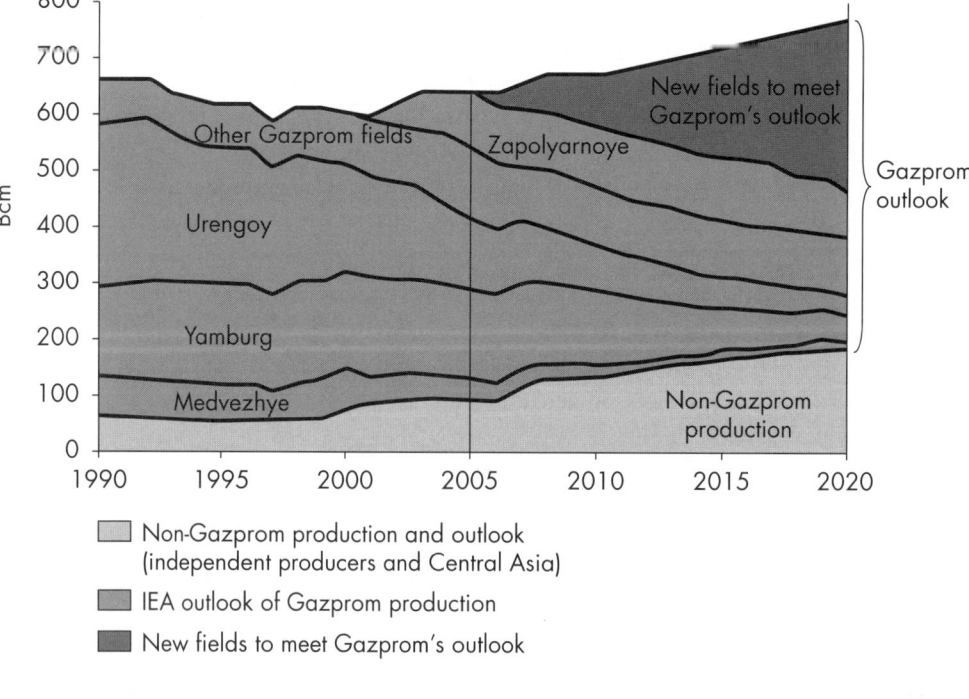

Non-Gazprom production and outlook
(independent producers and Central Asia)

IEA outlook of Gazprom production

New fields to meet Gazprom's outlook

Source: IEA estimates.

---

in Table 1, independents and Russian oil companies are keen to invest much more heavily in developing their gas resources which are often more accessible and more economic to produce, than in Gazprom's mega projects in increasingly difficult-to-develop regions.

Stemming domestic demand growth could also reduce the need for major new investments, depending on how active Russia is in enhancing energy efficiency and a more efficient use of its natural gas resources.

There are many uncertainties underlying security of supplies in the future and in particular in the next few years to Russian and traditional Russian export markets – Western, Central and Eastern Europe:

■ Will Gazprom attract timely investments to match the decline of its current fields and bring on stream enough new fields to ramp up new production?

■ Will Turkmen gas imports be negotiable (given higher Ukrainian negotiated prices)? More importantly, will Turkmenistan be able to ramp up production to 70-80 bcm quickly enough, given current production rates of about half this volume? What is the state of the Central Asian pipeline infrastructure and can it transport such a dramatic increase in gas volumes?

■ Will independent producers – Russian oil companies and independent gas producers – be able to ramp up production fast enough, if Turkmen gas is not forthcoming?

If not, what are the alternatives?

■ Can market reforms be implemented at a rapid enough pace to build the trust of independent producers, so that they make long-term investments in the sector to sustain and eventually increase gas production levels?

■ Would European gas exports be at risk – given Russia's unblemished long-standing supply record, save for the political cut-offs of gas supplies aimed at transit countries during negotiations over assets or tariff levels?

■ Can Russia enhance energy efficiency to compensate for *non-produced or non-imported* gas? Would this be through energy savings and a lower rate of growth of domestic gas demand? How politically palatable will this be as elections approach?

Difficult tensions in supply shift out beyond 2007, however, if the growth in domestic natural gas demand can be limited through enhanced energy-efficiency policies in Russia and more rational natural gas use. This is the case not only in terms of its consumption and transformation but also in terms of its production through investments to reduce fugitive emissions (leaks) along the natural gas transmission and distribution networks, to reduce the gas-fuel use to pump the gas along the transmission system by using more efficient compressor stations and through gas sector reforms to provide third party access to reduce the volumes of gas flared by oil companies.

Russian natural gas demand

Russian domestic gas demand contracted about 15% over 1992-98, reaching its lowest point in 1998. This is an extraordinary figure given the much deeper economic contraction during this period. Domestic gas demand in 2004 was 422 bcm, a growth of 11% since 1998, reaching 95% of 1992 levels. This reflects a growth rate of between 1-2% per year, except for 2003 when Russian natural gas demand grew by over 5% in that year alone. Figure 3 shows that heat and power generation account for almost 60% of total demand or 251 bcm in 2004. The share of natural gas in the fuel mix for electricity generation has remained relatively constant over the period 1992 to 2004, dropping from 45% to 44% of total. The overall share of natural gas in the thermal electricity generation fuel mix has remained constant over the decade ending in 2003 at between 65-66%, whereas the share of coal has increased from 23% to 29% as the share of fuel oil declined.

Total final consumption in 2004, the remaining 152 bcm of gas not used in transformation (into heat and power) or in the energy sector (13 bcm) or lost in distribution (6 bcm) was used in the various parts of the Russian economy. The industrial sector consumed 22% or 34 bcm in 2004, other sectors (mainly the commercial and public sector) another 37.5% or 57 bcm, another 13% or 19 bcm is used in the petro-chemical industry and pipeline transportation accounts for the remaining 27.5% or 42 bcm. As such, the volume of natural gas used within the

Figure 3   Russian gas demand by sector in 1992-2004

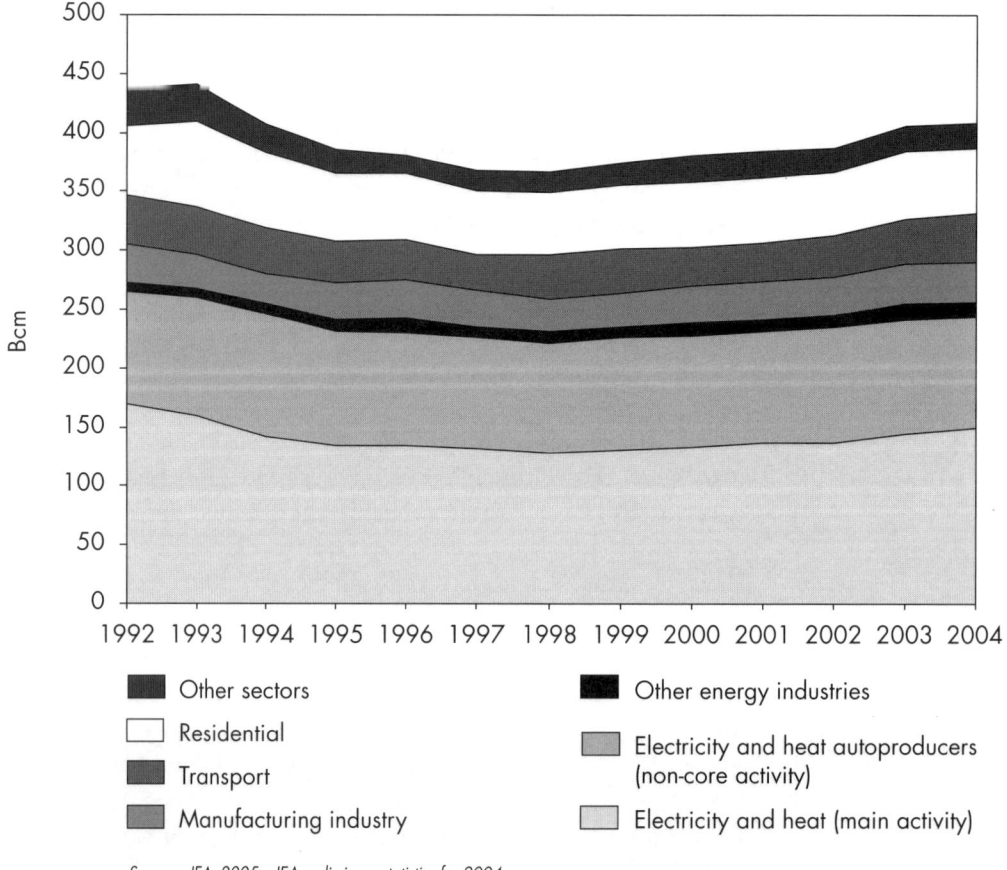

Sources: IEA, 2005c; IEA preliminary statistics for 2004.

Russian gas transmission sector – only to transport the gas – accounts for almost one third of total final consumption.

There are many areas of possible efficiency in Russian domestic gas demand. This includes efficiencies in the transmission and distribution process, the key focus of Chapters 3 and 4 of this book.

Russia's gas distribution sector

According to preliminary IEA statistics for 2004, total Russian gas demand was 422 bcm. This corresponds to Gazprom domestic supplies of 330 bcm plus independent gas production of 91 bcm in 2004 supplied to the domestic market.[11] Deducting the volume of gas consumed by Gazprom's transmission system, in the order of 42 bcm, about 380 bcm of gas flowed through high, medium and low-pressure distribution pipelines and was actually consumed in the domestic market in 2004.

Gasification of Russia's regions is not uniform, the western and central regions benefiting from the highest rates.[12] In 2005, on average 53% of Russian communities were supplied with natural gas. This is an increase from 42% in 1995 to 51% in 2002. In 2005 the divergence between urban and rural gasification rates was still very high, with 60% of the population in cities and towns being gasified while only 34% in rural communities. In 2004, Gazprom supplied gas to 80 000 communities (out of a total of 110 200), including 22.8 million homes (out of a total of 40.75 million), 12 200 industrial facilities and 29 600 boiler houses.

During the last years, growing retail prices for natural gas and the resolution for the most part of the non-payment problem have made the gas distribution, and more specifically the residential sector, more attractive for strategic investors. Gazprom is actively increasing its share in this promising market segment (see Figure 4). Since 1999, Gazprom has increased its control over gas distribution assets from a level of 13% to 75% of total in 2004. The Russian state controls the other 25% of gas distribution facilities, managed by Rosgazifikatsia. In late 2004, Gazprom regrouped its regional holdings (previously controlled by Gazprom's Regiongazholding) into a new company Gazpromregiongaz (Seleznev, 2004). However, at the time of completion of this study, no further developments on this front had occurred.

Gazprom is pushing for the maximum consolidation of Russian gas distribution companies. Since 2001, consolidation of gas distribution organisations has reduced their numbers from 318, to about 240 in 2005. Recent examples are the consolidation in Stavropol and Volgograd regions. In these regions several dozen gas distribution organisations were consolidated into one company. A similar consolidation is planned in the Krasnodar region, where more than 40 gas distribution organisations were functioning in 2004 (Seleznev, 2004). According to Gazprom, this consolidation is

11. According to Gazprom (Gazprom, 2005a), over half of the 330 bcm it supplied domestically was consumed by large industry such as the electricity and heat sector (46%), metallurgical (6%) and agro-chemical (7%) industry through high-pressure distribution pipelines. The medium and low-pressure distribution pipelines and service lines supplied the remaining volumes of gas to households (15%) and other consumers (26%).
12. The rate of gasification is determined by the share of apartments in a region having access to natural gas.

Figure 4   The structure of Russia's gas distribution sector

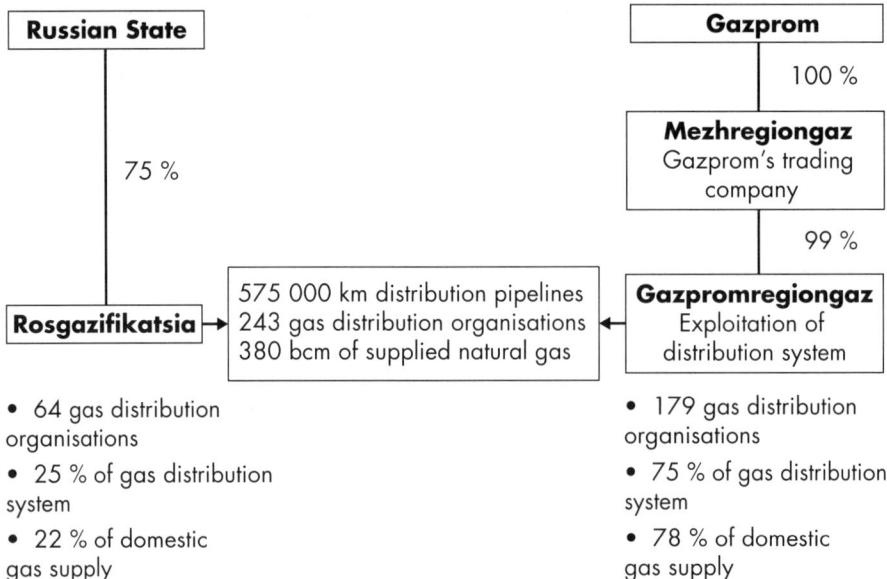

Gazpromregiongaz was created in 2004 by Mezhregiongaz (99%) and Lentranzgas (1%) in order to concentrate the management of gas distribution organisations controlled by Gazprom.

Sources: Gazprom, 2004a; Gazprom 2005a; Seleznev, 2004; Pravosudov, 2004b; IEA, 2005c.

the most effective way to create entities with the financial wherewithal to ensure the necessary investment in modernisation and expansion.

In mid-2004, Gazprom announced the unbundling of different operations in the structure of Gazpromregiongaz. The idea was to separate the functions of gas distribution (supply of gas and maintenance of the gas distribution systems) from the sales of the gas. Gazprom believes this separation of functions would improve transparency, provide for a more cost reflective tariff and support the modernisation of the low-pressure gas distribution system (Seleznev, 2004).

Gazprom is also actively participating in the gasification of Russian regions. Gazprom, in co-operation with a number of regional authorities, is continuing to expand the gas distribution system in order to increase the gasification of regions from 53% to 60% in 2008 and to increase the amount of rural gasification. In 2005-07, Gazprom's programme of gasification plans the construction of 12 000 km of distribution pipelines and to increase the use of existing capacity. The necessary investment is estimated at USD 1.3 billion. The investment for reconstruction and modernisation of existing facilities is estimated at only USD 36 million. This programme would require an additional 9 bcm/year of natural gas, of which 4.4 bcm to households and municipalities. Given the revenues generated by the current high price of natural gas on international markets, Gazprom is able to gasify at a faster rate, investing USD 1 billion in 2006 alone, given its higher revenues.[13]

---

13. The recent step up in Gazprom's regional gasification work also seems to be part of an agreement with the government that obliges Gazprom to spend at least USD 1 billion out of the USD 7 billion it should receive for the sale of 10.74% of its shares.

ENERGY EFFICIENCY: A HUGE POTENTIAL TO TAP

Energy efficiency is a key focus of the Russian Energy Strategy. Given the preponderance of natural gas in Russia's TPES, efficiency of natural gas production and consumption would be a logical focus for an efficiency effort. Greater gas efficiency would free up incremental gas for export, enhancing security of export supply. However, to date the implementation of energy-efficiency policies in Russia has not been especially successful.

The next part of this chapter presents the official Russian projections of overall energy-saving potential to 2020 qualified by our assessment of the existing barriers to effective implementation of energy-efficiency measures to achieve these goals in Russia as a whole and in its natural gas sector in particular. Russian ratification of the Kyoto Protocol could provide a new stimulus to overcome some of these barriers given the considerable environmental benefits in reduced GHG emissions that its energy-efficiency policies can also bring. Estimates are provided below of the overall volumes and potential for reduction of GHG emissions in Russia's gas transmission and distribution systems as well as the volume of gas flared in Russia.

Energy efficiency: necessary for sustainable growth

The Russian Energy Strategy estimates that Russia could reduce consumption of energy per unit of output by 40-50% from 2000 levels, but cost-reflective energy pricing will be needed to create the incentives to stimulate reductions in energy intensity. The Strategy projects that it would consume over 3 times more energy if it were to maintain its year 2000 energy intensity and still meet its year 2020 GDP growth target. In other words, the Strategy estimates that potential energy savings represent nearly two thirds of the additional energy needs to support its economic growth to 2020.

Figure 5 reflects the Strategy's outlook, distinguishing between the reduced energy demand due to structural changes in the Russian economy and that due to the implementation of specific energy-efficiency measures. The Strategy projects *structural changes* to account for about 70% of the reduction, as the economy shifts away from heavy industry and manufacturing to a more service-oriented GDP. Thus the Strategy expects the lion's share of improvements in energy intensity to happen naturally over time as GDP increases and restructures. *Technological changes* account for the remaining energy savings. The main technological potential of energy-saving measures is expected to be concentrated in the energy sector (including generation of electricity and heat) which could contribute to 36-40% of total savings. Industry's part in energy savings is estimated at 35-37% and residential 25-27%. Experience in IEA countries shows the importance of energy-efficiency policies and cost-reflective pricing to stimulate energy efficiency. The barriers to investments in energy efficiency, which Russia will need to surmount if it is to achieve its ambitious outlook are outlined below.

According to the Strategy, effective implementation of energy-efficiency measures would need to be supported by active economic reforms, including the rapid raising

Figure 5   Projections of Russian energy consumption to 2020

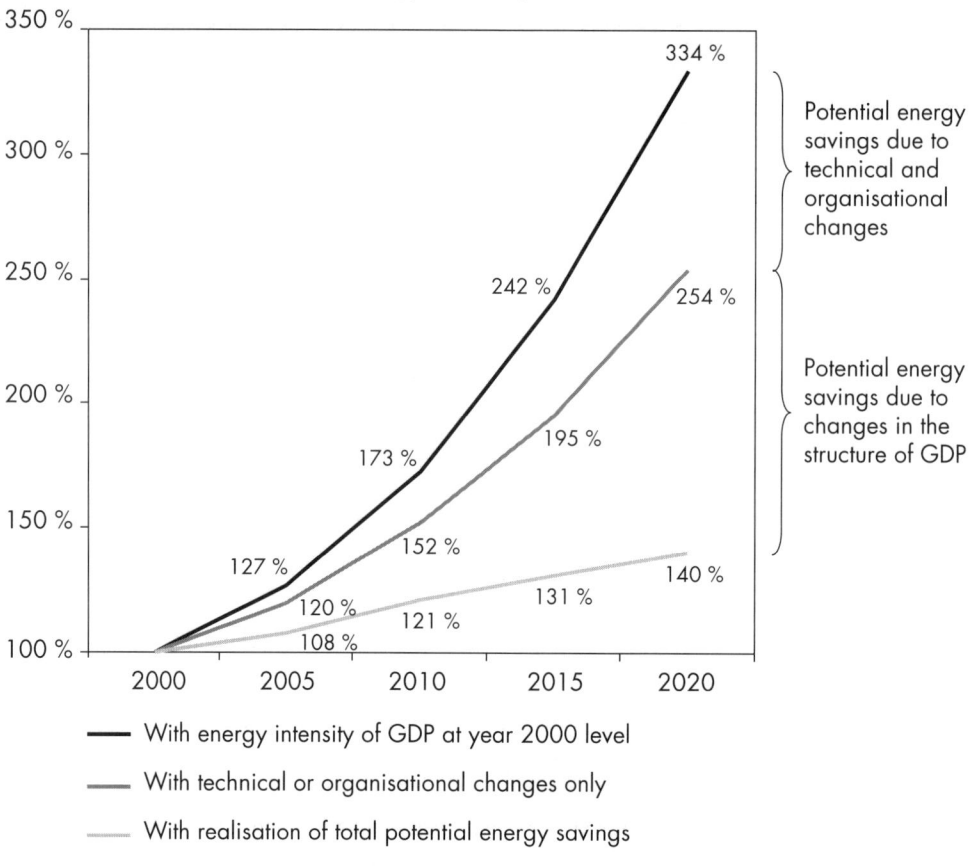

Source: Energy Strategy, 2003.

of energy prices to world levels and tariffs (including the elimination of the cross-sector subsidies in the electricity and gas tariffs). More efficient energy use would offset to a large extent the impact of gradually increasing prices on consumers. The creation of necessary conditions for competition in the current monopolistic gas and electricity markets is also considered a critical element (IEA, 2004c). In particular, such reforms would be essential to attract the necessary investments. Furthermore, efficiency and lower energy intensity generate environmental benefits through reduced GHG emissions and local air pollutants. To this extent, climate policy can play a key role in stimulating investments in this area.

General barriers to investments in energy efficiency

Under-investment in energy efficiency is not unique to Russia. Cost-effective energy-saving options are often neglected in other countries for various reasons.[14] Many factors other than direct, quantifiable costs affect consumer decisions. These include lack of information, technical, personnel and investment resources. Other barriers are uncertainty about energy prices, equipment performance and problems of equipment-

---

14. See IEA Web site for relevant IEA publications.

supply infrastructure. Then there is simple aversion to change. Most customers are interested in comfort, quality and availability as well as technological advances of more efficient appliances and technologies.

Efficiency-investment barriers more specific to Russia include low energy prices, monopolistic structures and lack of consumer control and metering equipment coupled with a system of billing (on a per resident basis) which provides little incentive for efficiency. On a more macro-economic level, the major barriers which continue to hamper investment include the lack of contract enforcement and an unstable investment environment. Mechanisms are needed to provide investors with greater stability and reduce the fiscal and legal risks of investments. A stability mechanism such as provided in production sharing agreements in the upstream oil sector or in the Energy Savings Company (ESCO) framework, could help minimise the risks of investing in energy efficiency.

A regional approach to energy efficiency is essential. The energy situation in each region depends on its natural resources, its distance from main distribution networks and its energy consumption patterns. Since 1995, an increasing number of regional administrations have developed legal, regulatory and institutional frameworks for energy efficiency.

Over the last five years, despite legislative interest in supporting and promoting investment in energy efficiency, few successes have emerged. Other factors hampering success on top of those already mentioned above include:

■ The small size of Russian energy-efficiency projects.

■ Lack of trained experts to develop bankable project proposals.

■ The outdated structure of building and residential energy-supply systems.

■ Lack of consumer-operated controls to regulate heating.

■ Lack of homeowner responsibility for repairs.

The Kyoto Protocol could enhance the attractiveness of some energy-efficiency investments through the use of its "flexible mechanisms".

Gas sector reforms: removing barriers to energy efficiency

As pointed out above, the overall efficiency of the gas sector in Russia is impeded in part by its monopolistic structure limiting upstream gas investments by independent gas producers and oil companies. More transparent and reliable third party access to both domestic and export markets would prove a major step forward. The Russian government's current attempts to promote this through various legislative initiatives and through efforts by the Anti-monopoly Service are welcome signs of an awareness. Unfortunately, this Service is grossly under-resourced. A more detailed discussion of the barriers to investments in energy efficiency throughout the various parts of the gas-supply chain is provided in the sectoral chapters of this book. However, the main common element throughout is the tight link between policies of the gas sector and overall energy reforms. Without proper policies, Russia will not tap its huge energy-saving potential along its gas-supply chain – including the better use of associated gas – in a timely way.

The best results from energy-efficiency programmes occur when they are directly embedded into sectoral policies and include measures that, together with raising national awareness of energy efficiency, both "push" the market (*e.g.* rational prices, mandatory efficiency requirement for equipment) and "pull" the market (*e.g.* incentives such as labeling). The incorporation of clear, feasible energy-efficiency measures and policies in the objectives of national climate policy could be another important driving force. A more structured institutional and legislative framework essential for the implementation of the Kyoto mechanisms in Russia could enhance the potential for energy-efficiency projects, at least in the medium term (see Chapter 2).

Incentives for energy efficiency via cost-reflective gas prices

Until recently, government regulation of domestic gas prices, at levels below market value, has been a major concern for the gas industry and its capacity to finance capital spending. Low domestic prices also affect prospects for stemming Russian gas demand growth and, therefore, incentives for energy efficiency, heightened competitiveness and increasing the amount of gas available for export. Raising domestic gas prices to market value is essential in reforming the gas sector and for the Russian economy as a whole (see Figure 6).

Figure 6   Historic and outlook for Russian domestic gas pricing

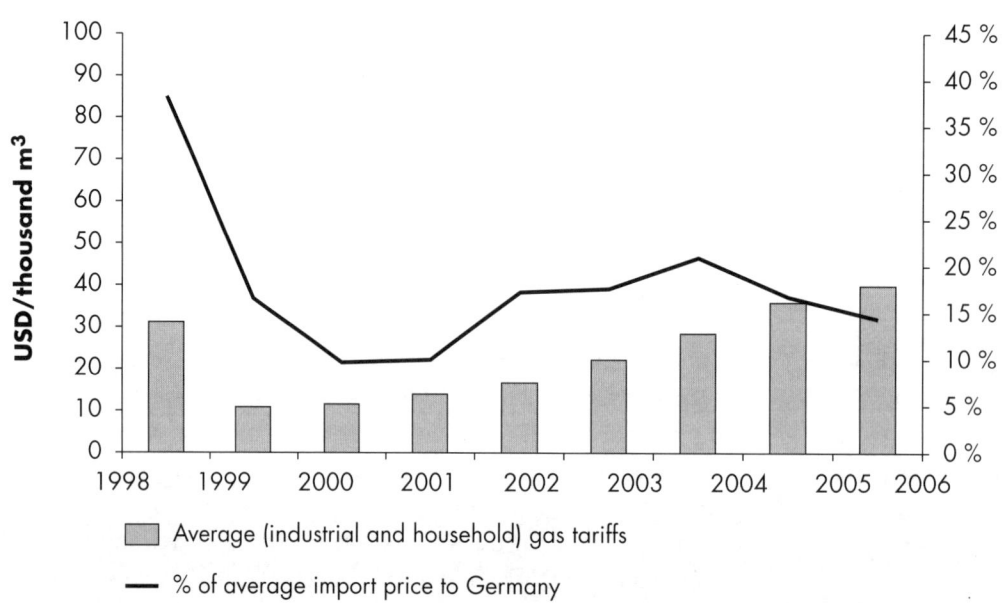

- Average (industrial and household) gas tariffs
- % of average import price to Germany

Sources: Ministry of Economic Development and Trade (MEDT); Energy Intelligence, 2006; Troika Dialog, 2006.

The Russian Energy Strategy envisages a continuing rapid increase in gas prices. Gas prices increased by about 70% on average in real local currency terms from the beginning of 2000 to the beginning 2004, including a 20% increase in 2004 and a further 23% increase in 2005. Tariffs are expected to increase by another 11% in 2006 and 8% in 2007. Unfortunately, prices are increasing on the low current base of USD 40/thousand m^3 whereas the same gas sells in Europe at USD 260/ thousand m^3 or (for now) costs USD 95/thousand m^3 at the Russian border.

The commitment to raise domestic gas prices has been institutionalised within the EU-Russia agreement signed in May 2004, where the EU gave its support for Russia's accession to the World Trade Organisation. The EU had argued that below-cost domestic tariffs represent a hidden trade subsidy. The Russian government promised to raise average gas prices to industry from USD 27/thousand m^3 in 2004 to between USD 37 and USD 42/thousand m^3 in 2006 and between USD 49 and USD 57/thousand m^3 in 2010, about the same levels as foreseen in the Energy Strategy.

GHG EMISSIONS IN RUSSIA: THE NATURAL GAS SECTOR AND GAS FLARING

In this book we examine GHG emissions from the Russian gas transmission and distribution systems, focussing on CO_2 emissions from gas combustion in compressors and fugitive CH_4 emissions due to venting from normal operations, as well as the unintentional CH_4 emissions from leaks and accidents.[15] We also examine GHG emissions from the flaring of associated gas by oil companies as they produce oil.

The two sub-sectors of the gas industry examined in more detail in Chapters 3 and 4 – transmission and distribution – account for the majority of total GHG emissions of the Russian gas industry. This is due to the scale of infrastructure in Russia covering great distances between production and consumption points, as well as to technical and physical characteristics of equipment such as age, optimisation parameters, energy efficiency and maintenance, etc.

According to the structure of Gazprom's energy consumption (see Figure 7), GHG emissions from production and processing account for only 10% of its energy consumption and are not large CO_2-emitters. A Gazprom and Ruhrgas study (Dedikov et al., 1999) showed that CH_4 emissions from production and processing in Russia are not significant at about 0.1% of gas production.[16] CH_4 emissions from storage should not be neglected and should be considered in further analysis of GHG emissions of the Russian gas industry when more information is available.[17]

---

15. Nitrous oxide (N_2O) is released to a much lesser extent, through combustion. However because it is negligible compared to emissions of CO_2 and CH_4, N_2O is not addressed in this study.
16. For details see Chapter 3 section "Methane emissions in Russia's gas transmission system" and Annex 5.
17. In Canada, CO_2 and CH_4 emissions from gas storage represent just over 1% of emissions of the gas transmission sector.

Figure 7   Gazprom energy consumption by sub-sector in 2000

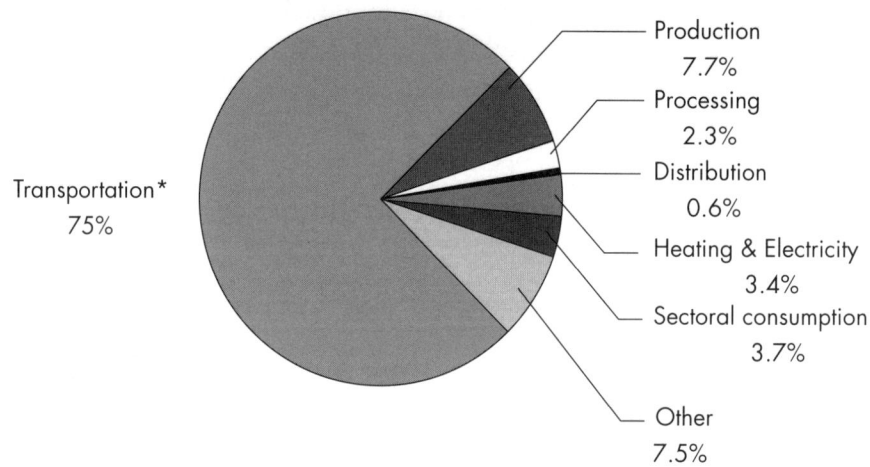

* includes underground storage of gas, which has relatively small energy consumption
Source: Gazprom, 2001a.

Gazprom's estimates of its GHG emissions from 1990 to 2000 and projections to 2012

Gazprom owns the entire Russian gas transmission system and controls three quarters of Russia's distribution network. As such its own estimate of its GHG emissions provides a useful starting point for evaluating these emissions for Russia as a whole.

Gazprom estimated its GHG emissions along the whole gas chain in 2000 at 232 MtCO$_2$ equivalent (CO$_2$e) including methane emissions in the order of 10 bcm (see Table 2). This represents a 12.4% decrease from 1990 levels. Energy related CO$_2$ emissions decreased by 20% over the period and Gazprom attributed this to the replacement of some of its most energy-intensive compressor units and the economic collapse. CH$_4$ emissions due to leaks decreased by only 9%.

Gazprom attributed the main energy savings in 2004 to its transmission sector, 84% of its total energy savings (Gazprom, 2004b). Gas production accounted for the remaining 14% of savings, and the processing, underground storage, drilling and capital repair of wells, 2%.

Table 2   Gazprom's estimates of its GHG emissions from 1990 to 2000 and projections to 2012

GHG emissions from gas extraction to distribution	1990	2000	2008	2012	Emission reductions 2000-12
Emissions of CH$_4$, bcm	11.0	10.0	7.8	5.1	4.9
Emissions of CH$_4$, MtCO$_2$e	160.7	148.0	114.6	75.0	73.0
Emissions of CO$_2$ from gas combustion, MtCO$_2$e	105.4	84.0	74.0	70.0	14.0
Total, MtCO$_2$e	**266.1**	**232.0**	**188.6**	**145.0**	**87.0**
Share of total Russian GHG emissions	8.7%	12.4%	-	-	-

Sources: Silva et al. (2004), using VNIIGAZ information; Vernadskiy Foundation, 2004.

As shown in Table 2, Gazprom projects a faster downward trend of its GHG emissions in the future with total emissions in 2008 falling to 71% of 1990 levels. This would bring its emissions to almost half (54%) its 1990 levels by 2012. The investment programme tied to these emission reductions are discussed in Chapter 3. Gazprom evaluates its total annual potential for GHG emission reductions at 87 $MtCO_2e$ between 2000 and 2012. This is equivalent to roughly 6% of Russia's total CO_2 emissions in 2004. However, an independent estimate of Gazprom's 2002 GHG emissions (Mielke *et al.*, 2004) showed an increase over 2000 levels. They amounted to 237 $MtCO_2e$, of which over 150 $MtCO_2e$ are CH_4 emissions. Although modest increases, these latest estimates for 2002 reflect a trend completely at odds with Gazprom's outlook for 2008 and 2012. However, it could be misleading to compare data from different sources and draw conclusions – especially from one data point.

Table 3 presents Gazprom's projections of GHG emissions by sub-sector for the period 2000 to 2012.[18] CH_4 is expected to be the main source of total GHG emission reductions (*i.e.* 72 of the 86 $MtCO_2e$) due to a reduction in gas leaks and losses of 4.9 bcm. Energy-efficiency measures at compressor stations are expected to reduce the gas consumed in the transmission system by about 7.2 bcm and thus contribute to the reduction of 13.2 $MtCO_2e$.

According to Gazprom's estimates, gas transmission will continue to provide the largest opportunity for GHG reductions. Table 3 projects reductions of only about 15 $MtCO_2e$ for each of the gas production and distribution sectors. As discussed in more detail in Chapter 4, the values presented below for the distribution sector should be considered as underestimations given that Gazprom had a much smaller share of control of the distribution network at the time these projections were made.

Table 3　　　Gazprom's potential GHG emission reductions in 2000-12

Sub-sector	Reduction of CH_4 emissions		Reduction of gas consumption for combustion needs		Total GHG reductions	
	bcm	$MtCO_2e$	bcm	$MtCO_2e$	bcm	$MtCO_2e$
Transportation	2.6	38.2	7.2	13.2	9.8	51.4
Underground storage	0.3	3.7	-	-	0.3	3.7
Production	1.0	14.7	0.4	0.8	1.4	15.5
Processing	0.1	0.7	0.2	0.3	0.2	1.0
Distribution	1.0	14.7	-	-	1.0	14.7
Total	**4.9**	**72.0**	**7.8**	**14.3**	**12.7**	**86.3**

Source: Energy Security of Russia, 2005.

---

18. The sectoral potentials for energy savings are discussed in more detail in sectoral chapters of this book.

This estimated potential for GHG emission reductions in Russia's gas sector is based largely on the implementation of regular refurbishment measures to maintain and increase Gazprom's operational capacity in an economically attractive way. The expected progressive increase in Russia's domestic gas prices, along with the expected increase in production costs, should further improve the attractiveness of these efficiency measures through the enhanced value of gas savings and other operational benefits. In the sectoral chapters of this book we assess the potential for climate policy to enhance the rate of refurbishment and modernisation beyond current practices.

Estimates of GHG emissions in gas transportation, distribution, and gas flaring

Estimates of GHG emissions from Russia's natural gas sector are highly uncertain. This is because there is currently no complete inventory of historical and current emissions; and Russia-specific emission factors for GHG emissions for various gas-sector activities still have to be developed.[19] Figure 8 shows our assessment of the main sources of GHG emissions for 2004 in the sub-sectors of the natural gas industry examined in this study. The estimates are based on a bottom-up extrapolation of available GHG emission data from various studies undertaken by international experts in co-operation with Gazprom and on information provided by Gazprom.[20] These estimates should therefore be viewed as orders of magnitude for Russia's emissions and not precise estimates.

The shaded columns show natural gas consumption and leaks, and the black columns – the corresponding GHG emissions. Methane is converted into CO_2 equivalent based on its global warming potential equal to 21 (IPCC, 2000). This explains the much larger impact of CH_4 leaks in terms of greenhouse gas effect than the CO_2 emissions due to combustion of fuel-gas or associated gas.

The first two series of columns represent CH_4 emissions due to gas losses at transmission and distribution systems from leaking components during normal operations, maintenance and repair work and due to accidents. The remaining two series reflect CO_2 emissions from gas combustion in gas turbines at compressor stations, as well as from flaring of associated gas at oil producing wells.

In terms of natural gas, the four shaded columns combined represent just over one third of Russia's gas exports to OECD Europe in 2004. It is important to note however, that about 60% or 41 bcm of this gas is consumed by compressors to maintain the pressure along the gas transmission system. While there is a huge potential for enhancing the efficiency of these compressors, it is impossible to reduce this volume below a level reflecting the best available international compressor technology.

---

19. As we highlight throughout the book, the emission factors proposed by IPCC (1996, 2000) are based largely on North American studies and cannot be directly applied to the Russian gas sector.
20. IEA statistics do not include CH_4 emission data, only CO_2 emission data.

Figure 8   Estimated structure of GHG emissions from Russian gas transmission and distribution systems and gas flaring in 2004

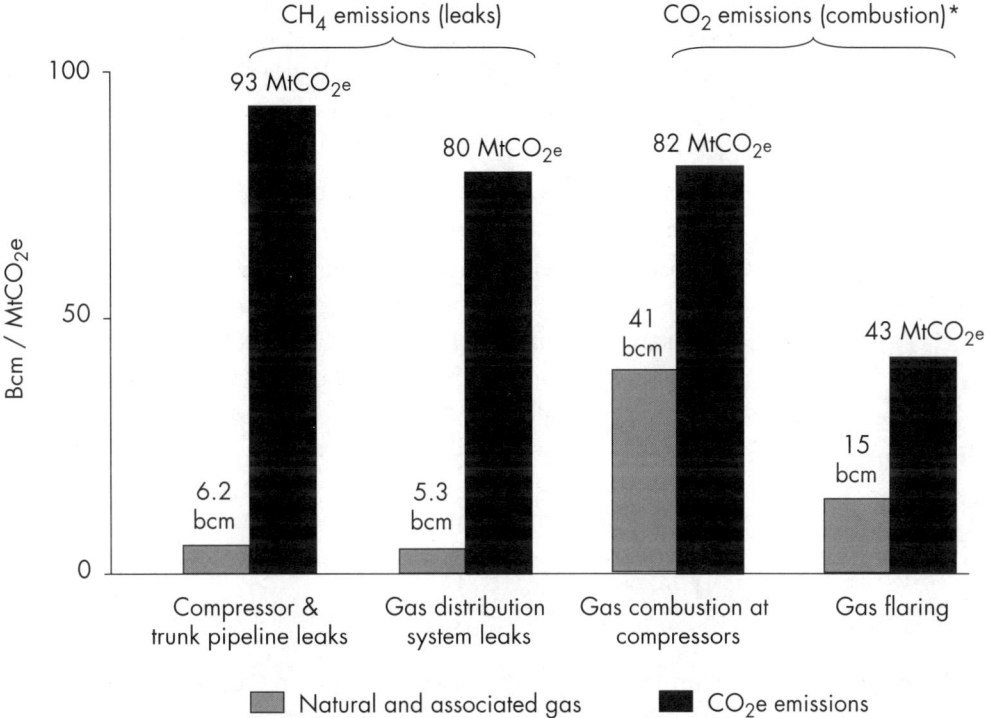

* The GHG emissions from flaring of associated gas are essentially CO_2, but include also a part of the unburned gas in the form of CH_4 emissions, as shown in Annex 1.

Source: The estimates are based on a bottom-up extrapolation of available GHG emission data from various studies and on Gazprom's information. Details are provided in the sectoral chapters of this study.

Conservative conversion factors are used to calculate the emissions in CO_2-equivalent (for details see Annex 1):

■ 15 $kgCO_2e/m^3$ is used for the conversion of 1 m^3 of natural gas directly released into the atmosphere through leaks and losses and accidents.

■ 2 $kgCO_2e/m^3$ is used to convert 1 m^3 of natural gas used at compressors along the transmission system.

■ 2.9 $kgCO_2e/m^3$ is used to convert 1 m^3 of associated gas flared at the wellhead as oil is produced.[21]

Our estimates show that CH_4 emissions from leaks along both the transmission and distribution systems are in the order of 170 $MtCO_2e$ and account for over 60% of GHG emissions of the three sectors examined in this study.

---

21. The different conversion factor for associated gas is higher because this gas is heavier than natural gas and part of this gas is directly vented into the atmosphere.

Estimates of CH_4 emissions in the distribution network reflect the highest level of uncertainty given the lack of available data on the current stock of gas distribution facilities and the absence of comprehensive measurement programmes. These estimates are calculated using the average emission rate of 3.2% of distributed gas, based on assessments by Russian experts, including Gazprom (see Chapter 4).

CO_2 emissions from the gas transmission system are estimated at 82 $MtCO_2$. The consumption of fuel gas by compressor stations amounts to approximately 6% of the gas throughput volumes of 687 bcm in 2004 or about 41 bcm. Compressor stations represent 27% of GHG emissions of the three sectors examined in this book.

Gas flaring represents another significant source of CO_2 emissions, totalling about 43 $MtCO_2$ or 14% of total GHG emissions of the three sectors studied. In terms of natural gas, the volume of flared gas represents 14.7 bcm or 2% of total Russian gas production in 2004. This increased slightly to 15.0 bcm in 2005 (Energy Sector of Russia, 2004, 2005).[22]

Overall, emissions from gas-related activities totalled about 300 $MtCO_2e$. This represents roughly 15.3% of the estimated total Russian GHG emissions in 2004.

Estimates of potential reductions in gas losses and GHG emissions

Figure ES-1 in the Executive Summary reflects the IEA's estimates of potential reductions in emissions and combustion of natural gas in the various sectors. Over 60% of GHG emissions can be reduced along pipeline networks (over 100 $MtCO_2e$). A larger share of the potential reductions of CH_4 leaks can be found in the distribution sector as opposed to the transmission system. In terms of natural gas savings, however, these estimates are much less significant (6 bcm) than those which can potentially be reduced at compressor stations along the transmission system (8 bcm) or flared by oil companies (15 bcm).

In terms of orders of magnitude, the distribution sector clearly holds a potential similar and possibly greater than the transmission sector in terms of reductions of CH_4 leaks. The study results point to the greater potential in the transmission sector to reduce natural gas use at compressor stations through the use of more efficient best-available-technologies (BAT). Gazprom itself projects savings through reduced compressor consumption of up to 10 bcm by 2010. Kyoto-related investments could speed up this process.

Study results related to the gas distribution network encompass the most uncertainty, for the reasons outlined earlier. More independent studies are also necessary in the transmission sector which would be seen as an objective estimate for GHG emission inventories, in particular concerning CH_4 emissions.

---

22. It should be noted that experts consider official reports by oil companies often underestimate volumes of flared gas. Experts in Russia believe actual volumes could be more than double official estimates.

The potential to completely reduce the volume of gas flared has been assumed despite our understanding that not all currently flared gas can be economically used. This assumption reflects more a need for enhanced transparency related to flaring activity in Russia and globally, as well as the current initiative of the Russian government to raise the costs of gas flaring and provide more incentives to stop this activity. It also reflects the IEA's view of the need for more transparent and reliable third party access to Gazprom's transmission infrastructure and of the benefits this could have to enhance gas supply security and reduce its negative impact on the environment.

II. RUSSIAN CLIMATE POLICY FRAMEWORK

RUSSIA IN THE CONTEXT OF THE INTERNATIONAL CLIMATE POLICY

Estimates and trends of Russian GHG emissions from 1990 to 2004

Russia's economy contracted sharply during 1990-98 after the break-up of the Soviet Union. This resulted in a massive reduction of Russia's CO_2 energy-related emissions. Even so, in 2003 Russia was still the world's third largest CO_2 emitter, after the United States and China (see Figure 9).

Figure 9

World energy-related CO_2 emissions in 2003, by country (25.2 $GtCO_2$)

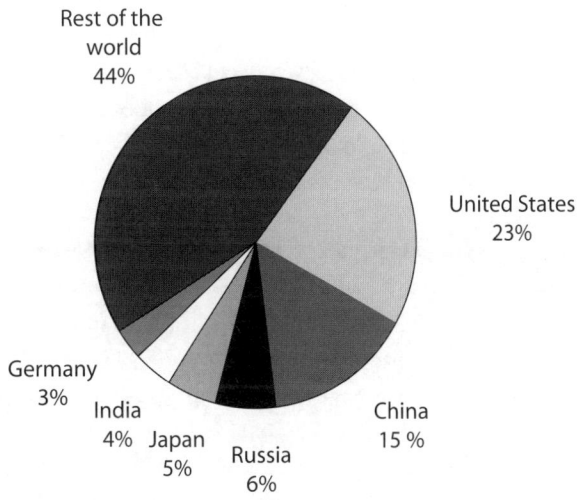

Source: IEA, 2005b.

Russia's GHG emissions and its Kyoto commitment

Table 4 provides the latest Russian official estimates of its GHG emissions including CH_4 from the country's Third National Communication to the Secretariat of UNFCCC (United Nations Framework Convention on Climate Change) (ICCC, 2002). According to this official source, Russia's 1999 total GHG emissions amounted to 1 880 $MtCO_2e$, representing a 38% drop compared to its 1990 level. The UNFCCC Report on the In-depth Review of the Third National Communication of Russia (Silva *et al.*, 2004) points out that these estimates are highly uncertain largely due to the difficulties in collecting the data on fugitive methane emissions and the absence of specific emission factors for Russia (ICCC, 2002).

GHG emission levels for 1990 are also uncertain, an important point as they determine Russia's emissions target under the Kyoto Protocol, the so-called Assigned Amount of GHG emissions. As Russia's commitment under the Kyoto Protocol is to stabilise its emissions at 1990 levels, its total GHG emissions must not surpass this level during the five year period from 2008 to 2012.

Table 4   Russia's estimated total GHG emissions

GHG emissions, MtCO$_2$e	1990	1995	1999 estimates	% change 1999/1990
CO$_2$	2 360	1 590	1 510	-36%
Energy-related CO$_2$*	*2 320*	*1 570*	*1 470*	*-37%*
CH$_4$	550	390	290	-47%
Energy-related CH$_4$	*401*	*278*	*199*	*-51%*
N$_2$O	98	43	35	-64%
PFC, HFC, SF6	40	38	42	5%
Total	**3 050**	**2 060**	**1 880**	**-38%**

* Includes losses and fugitive emissions: associated gas flaring and emissions from coal mining.
Source: ICCC, 2002.

The Russian Third National Communication estimated the 1990 level of total Russian GHG emissions at 3 048 MtCO$_2$e, with CO$_2$ emissions amounting to 2 360 MtCO$_2$. The IEA estimates that Russia's energy-related CO$_2$ emissions amounted to 2 023 MtCO$_2$ in 1990.[23] The 337 MtCO$_2$ difference between Russia's official figure for CO$_2$ emissions (see Table 4) and the IEA's sector-based CO$_2$ emissions inventory shows the importance of a comprehensive system of emission estimates and inventory procedures for a country the size of Russia, where the collection of energy statistics is difficult.[24] There is even more uncertainty concerning the statistics of energy losses and CH$_4$ leaks, which also need to be estimated to determine compliance to the Kyoto commitments.

An independent report on Russia's 1990 GHG emissions estimates combined CO$_2$ and CH$_4$ to be 12% lower than reported in the country's Third National Communication, with CH$_4$ emissions accounting for most of the difference (CENEf and PNNL, 2004). This discrepancy highlights the uncertainty of CH$_4$ emissions, stemming from uncertainty on emissions in the oil and gas sector, and coal mining.

Russia's apparent surplus in Assigned Amount Units (*i.e.* actual emissions below its commitment) is much larger than originally expected, the result of the severity of the economic crisis over the 1990s and the ensuing decline in energy consumption. While further work on emission inventories is clearly needed and uncertainties exist, it is still useful to assess the size of Russia's potential emissions surplus and its potential to further reduce GHG emissions.

A significant surplus under the Kyoto Protocol

The Third National Communication estimates the total Russian GHG emissions in 1999 at 1 880 MtCO$_2$e, or 61.5% of the Russian 1990 GHG emissions.[25] In 1999, CO$_2$ accounted for 80% of the total or 1 510 MtCO$_2$. CH$_4$ with 15.5% of the country's total emissions is the second largest GHG in Russia.

---

23. The IEA calculates and estimates energy-related CO$_2$ emissions, but not CH$_4$ emissions.
24. This difference can also be due to the use of different approaches for estimating emissions (*i.e.* sector-based or reference) and difficulties in disaggregating Former Soviet Union (FSU) data.
25. The year 1999 is the last year for which emission calculations are provided in the Third National Communication (ICCC, 2002).

According to the latest preliminary IEA data, Russia's energy-related CO_2 emissions in 2004 represented about 1 529 $MtCO_2$, increasing by 7% from its lowest emission level in 1998. However, other greenhouse gases, in particular energy-related CH_4 emissions from the oil and gas industry, represent a significant share of the Russian GHG emissions. For this reason, a factor of 1.25 is used to estimate total Russian GHG emissions to around 1 910 $MtCO_2$e in 2004.[26] With 1 140 $MtCO_2$e below its annual assigned amount at present, Russia is expected to have a significant surplus that it could either sell to other Kyoto Parties, or bank for future use.

Main drivers of the evolution of GHG emissions

The evolution of Russian GHG emissions over the 1990s (see Figure 10) reflects the sharp contraction of the Russian economy during its market transition. The observed growth since 1999 is due to Russia's strong economic recovery, stimulated by the increase in world energy prices. Russia's economy remains CO_2-intensive with 1.18 $kgCO_2$ per unit of GDP, more than 2.5 times higher than the OECD average (computed on a 2000 USD purchasing power parity (PPP) basis). Canada, whose geography and natural resources are comparable to those of Russia, has a carbon intensity of 0.6 $kgCO_2$/USD – less than half of Russia's level.[27] However, IEA statistics show a reduction of Russia's energy intensity of GDP of about 3% per year from 1998 to 2001 and of about 5% since 2002.

Figure 10   Russian GDP, energy consumption and CO_2 emissions, 1990-2004

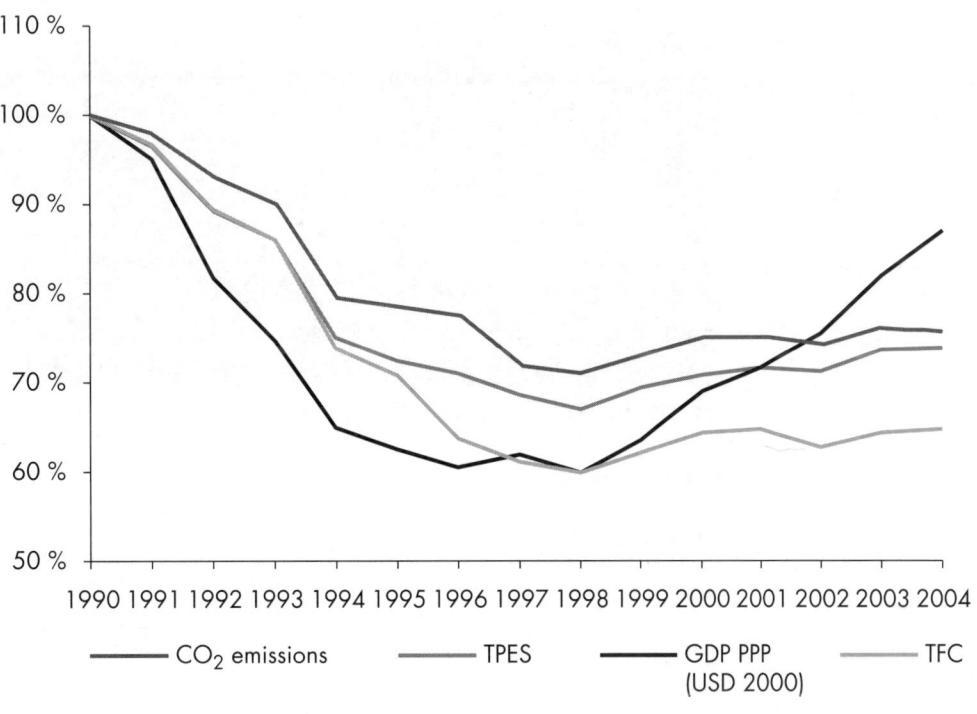

Sources: IEA, 2005b; preliminary IEA data for 2004.

---

26. As in the Report on the In-depth Review (Silva *et al.*, 2004), considering the share of CH_4 in total GHG emissions in Russia, the conservative ratio 1.25 proposed by Haites (2004) is used to estimate total Russian GHG emissions.
27. This comparison should be used with caution, taking into account the different structure of the economies of these two countries such as the different travelling distances, size of homes and structure of manufacturing.

Russian experts attribute this drop to structural changes. However, World Bank analysis concludes that the service sector's share in Russia's GDP can be overestimated in an effort to reduce oil revenues through trading companies to lower effective tax rates. The Russian estimates (Kokorin *et al.*, 2004) are similar to IEA data showing the drop in CO2 content of Russian GDP at more than 4% per year, attributed mainly to the rapid increase in oil and gas export revenues.28

CO_2 and CH_4 emissions by sector

Figure 11 presents the estimates of Russia's energy-related CO_2 emissions by sector, based on IEA methodology (IEA, 2005b).[29] In 1999, the oil and gas sector accounted for about 60% of Russia's CH_4 emissions, estimated at 166 $MtCO_2e$ or 9% of total GHG emissions (ICCC, 2002).[30] These emissions originate from the extraction, transportation and distribution of oil and gas.[31]

Figure 11 Sectoral structure of Russian CO_2 emissions from energy combustion

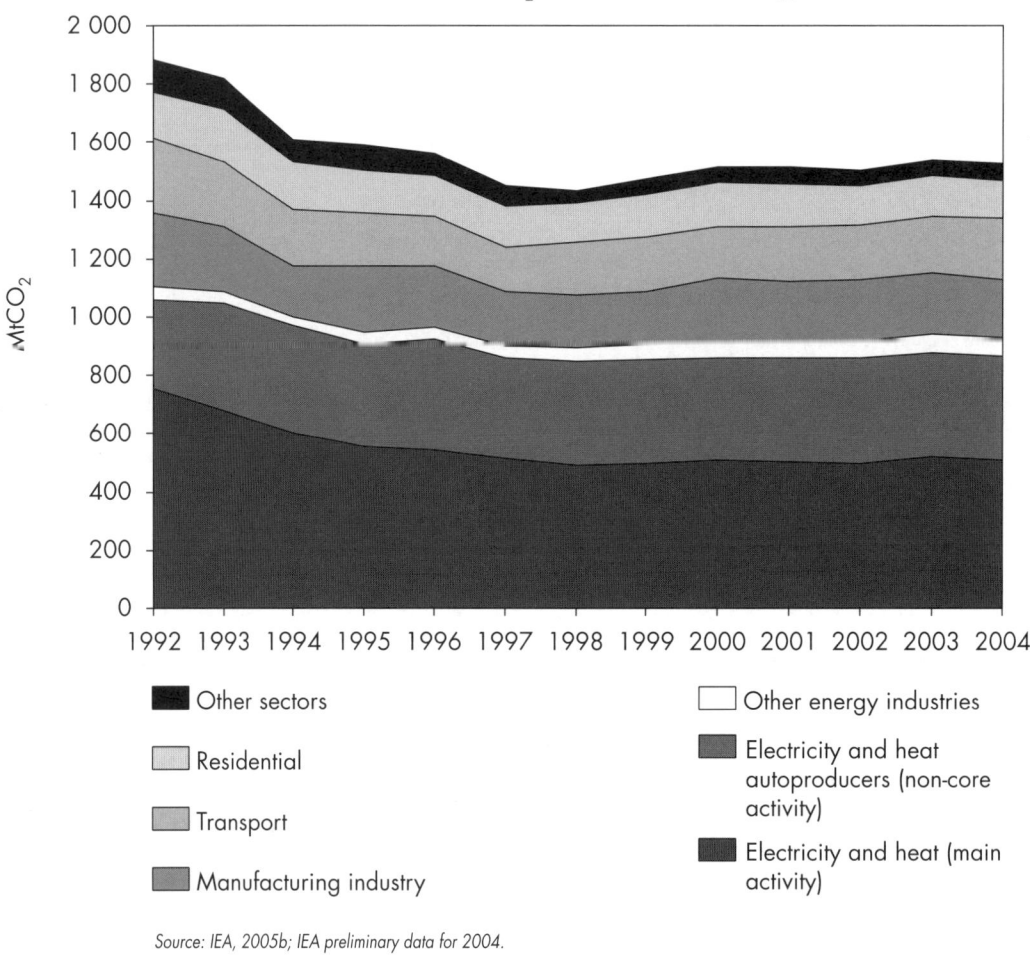

Source: IEA, 2005b; IEA preliminary data for 2004.

---

28. Energy and CO_2 intensity of Russia's GDP in 2004 were 13-15% below 1990 levels.
29. It should be noted that differences exist between IEA and IPCC guideline breakdown of sectors.
30. The other 40% of CH_4 emissions are generated essentially by the coal mining sector, agriculture and waste.
31. For 2001, CENEf and PNNL (2004) estimate Russia's overall fugitive methane emissions at 172 $MtCO_2e$, also largely coming from the natural gas sector.

The structure of Russian CO_2 emissions remained stable over 1992-2004. Electricity and heat generation accounted for about 60% of total CO_2 emissions. The manufacturing and transport sector emissions remained stable, each accounting for about 13-14% of CO_2 emissions, while the residential sector accounted for 9%.

Projections of Russian GHG emissions

The Third National Communication's projection is provided only for the energy-related CO_2 emissions and assumes a CO_2 intensity of energy consumption fixed at the year 2000 level for all scenarios (see Table 5).[32] The share of natural gas in the energy mix is projected to fall from 48% to 42-45% while the share of oil remains stable. The share of coal will rise slightly from 20% to 21-23%. Nuclear power will attain about 6% of total primary energy supply and renewables 1.1-1.6%. Variations in future CO_2 emissions between the three scenarios are entirely driven by assumptions on economic growth and on projected changes in the energy intensity of GDP.

The Optimistic (NC I) and Pessimistic (NC II) Scenarios are based on government programmes for economic and social development, and on the objectives of the Russian Energy Strategy (2002 version). The Realistic Scenario (NC III) combines a moderate GDP growth and rather moderate achievements in energy-efficiency improvements.

The annual rates of reduction of energy intensity are 3.7%, 2.5% and 2.0% for Scenarios I, II, and III respectively, within the range of observed rates in Russia and other countries with economies in transition between 2000 and 2003 (average of 3-4%). The Russian Energy Strategy (2003), assumes that structural changes would contribute to two thirds of the reduction in projected energy intensity. The remainder would come from tapping the large energy-efficiency potential, in particular by refurbishing obsolete energy-intensive equipment in the energy sector and efficiency improvements in various end-uses (see Figure 5).

All three scenarios show Russia fulfilling its 2008-12 Kyoto commitments (see Figure 12). In 2010, Russia's CO_2 emissions could reach 75-89% of 1990 levels and only surpass them by 2015 in the Realistic Scenario. By 2020, Russia's CO_2 emissions would reach 81 to 114% of the 1990 level.

Table 5   Russian projections of economic growth and CO_2 emissions for 2000-12

Macroeconomic indicators (annual rates of growth)	Scenario I Optimistic	Scenario II Pessimistic	Scenario III Realistic
GDP	5.2%	3.3%	4.5%
Energy intensity of GDP	-3.7%	-2.5%	-2.0%
Energy consumption	1.5%	0.8%	2.5%
CO_2 emissions	1.5%	0.8%	2.5%

Source: ICCC, 2002.

32. Given that the CO_2 emissions represent about 80% of Russia's total GHG emissions, this projection sufficiently reflects the major GHG emission trends in Russia. However, as it does not include CH_4 emissions nor CO_2 emissions from changes in land-use and forestry activities, this projection includes significant uncertainties. A more comprehensive national level analysis should be performed to address these uncertainties.

The 2003 Russian Energy Strategy confirms these projections, with energy-related GHG emissions reaching 75-80% of the 1990 level in 2010. In this Strategy, emissions would remain lower than their 1990 level until after 2020 (Energy Strategy, 2003).

A more recent analysis by the Ministry of Economic Development and Trade (MEDT) based on the scenarios of the Mid-term Programme of Social and Economic Development (2006-2008), projects the growth of CO_2 emissions in the same range - at 1% to 2.5% per year (MEDT, 2006).[33] The IEA *2004 World Energy Outlook* (WEO) estimates Russian CO_2 emissions in 2010 at 76% of the estimated 1990 level (IEA, 2004b).

Table 6 presents the surplus Assigned Amount Units (AAUs) for Russia on an annual basis ranging between 262 $MtCO_2$ and 638 $MtCO_2$ for energy-related CO_2. For all GHG, the volume of Russia's surplus AAUs is estimated between 330 and 800 $MtCO_2e$.

These estimates confirm Russia as a potentially major player in the market for GHG emission reductions under the Kyoto Protocol. With adequate domestic implementation, climate policy could become an important component of Russia's energy picture, all the more so as the sector reveals important potentials for cost-effective energy-efficiency improvements in a range of industrial activities and end-use sectors.

Figure 12   Projections of CO_2 emissions in Russia relative to 1990 levels

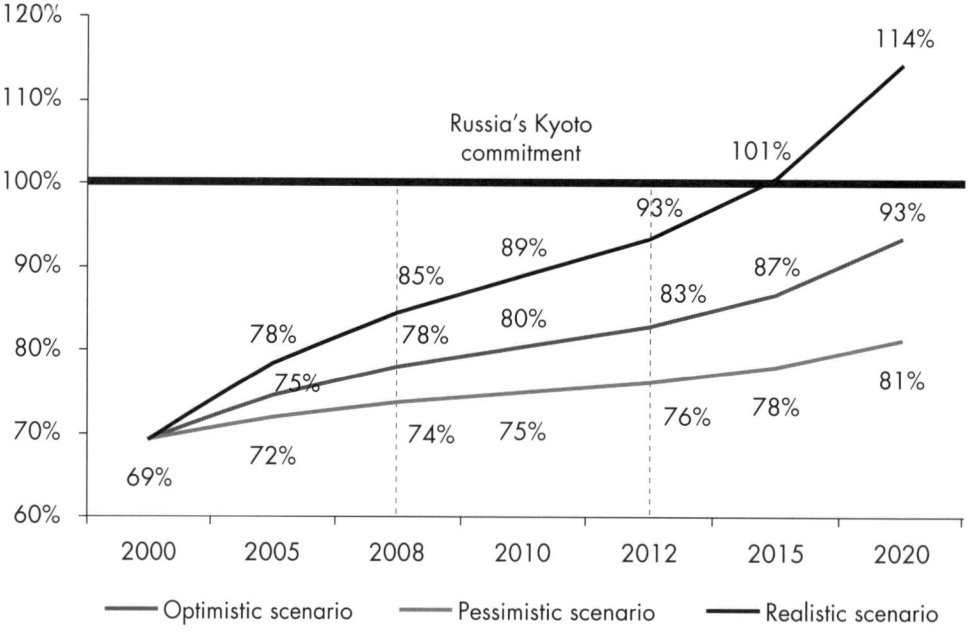

Source: ICCC, 2002.

---

33. The business-as-usual and innovation-based scenarios assume an annual GDP growth rate of 5% and 6.3-6.5%, respectively.

Table 6   Russia's estimated surplus of Assigned Amount Units in 2010

Scenario	Russia's estimated AAUs surplus in 2010	
	CO$_2$ emissions from fuel combustion, MtCO$_2$	All GHG emissions, MtCO$_2$e
3NC Optimistic	463	579
3NC Unfavourable	591	738
3NC Realistic	262	328
ES2003 Probable	520	650
ES2003 Optimistic	638	797
WEO Reference (2004)	490	612

Sources: based on the ICCC, 2002; Silva et al., 2004; IEA, 2004b.

THE KYOTO PROTOCOL AND RUSSIA: MANAGING THE SURPLUS

Given Russia's considerable surplus and its technical potential for further GHG emission reductions in the energy sector, the Kyoto Protocol flexibility mechanisms allow Russia to potentially become a central actor for the least-cost achievement of Annex I countries' commitments – as many of them are projected to have difficulties reaching their Kyoto commitments through domestic means only.

Figure 13 summarises existing and possible mechanisms for international transactions of Kyoto units to 2012, the end of the first commitment period of the Kyoto Protocol. This reflects the uncertainty concerning the possible characteristics of a future international climate regime.

Figure 13   Existing and proposed mechanisms for international emissions transactions under the Kyoto Protocol

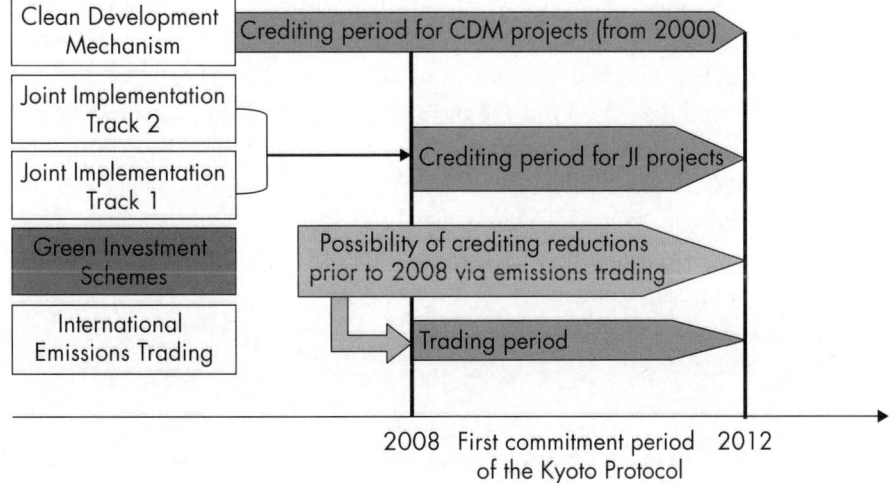

The Kyoto Protocol established two project-based flexibility mechanisms, the Clean Development Mechanism (CDM) and Joint Implementation (JI) that allow countries to acquire emission reductions from projects undertaken elsewhere. The CDM covers projects in developing countries. JI is open to industrialised countries in Annex I, and is a likely vehicle for the transfer of emission reductions from Russia to countries exceeding their Kyoto commitments, if Russia chooses to do so and meets all its eligibility requirements. JI includes two tracks to allow participation by countries not yet fully eligible for the Kyoto flexibility mechanisms (see Box 1). Emissions trading, the third flexibility mechanism, allows countries with commitments under Kyoto to trade AAUs.

The fourth mechanism – Green Investment Scheme (GIS) – is not part of the Kyoto Protocol. The GIS was proposed by Russia and other transition countries to "green" surplus AAUs – originally dubbed as "hot air" by environmental non-governmental organisations. In fact, GIS should link the transfer of surplus AAUs from transition economies to specific GHG mitigation activities (see Box 2). The "greening" should become a solution for many governments who are reluctant to purchase surplus AAUs which are not perceived as legitimate GHG emission reductions but more as a by-product of the economic collapse of countries with economies in transition during 1990s.

International Emissions Trading under Kyoto: a market outlook

In the context of Annex I countries, the demand side of the potential emissions trading market can be estimated using the IEA *2004 World Energy Outlook* (IEA, 2004b). The *World Energy Outlook* (WEO) Reference Scenario projects energy needs and related CO_2 emissions for regional groups including the 25 countries of the European Union, other OECD regions, and countries with economies in transition (EITs). The WEO indicates an upper bound of trading needs across these regions: it assumes the introduction of no new policy to reduce emissions beyond those that have been implemented as of July 2004. The IEA's more recent publication on emissions trading (IEA, 2005a) combines the WEO's projected trend in energy-related CO_2 with other GHG emissions including removals by sinks to evaluate countries' compliance.

The WEO's regional analysis suggests a demand by Annex I countries – excluding the United States and Australia which withdrew from the Kyoto Protocol – of some 840 $MtCO_2$e in 2010. Once eligible for emissions trading, the economies in transition (EITs) could be in a position to sell some 1 190 $MtCO_2$e of AAUs that year. This would leave the Kyoto regime with excess emissions of about 350 $MtCO_2$e, assuming that EITs transfer their excess allowances in full to other countries. Under the 2004 WEO's Alternative Policy Scenario, the projections indicate a higher excess of allowances totalling approximately 530 $MtCO_2$e across all industrialised countries.[34] Excluded from the WEO's calculations, the CDM will augment the potential supply of units for compliance with Kyoto commitments.

---

34. In the WEO 2004 Alternative Policy Scenario, buying regions could demand an average of 700 $MtCO_2$e each year, while potential sellers could offer 1 235 $MtCO_2$e for transfer.

Box 1   Joint Implementation (JI): two tracks

Joint Implementation (Kyoto Protocol's Article 6) is the project-based flexibility mechanism introduced to tap the low-cost potential of GHG mitigation in countries with economies in transition (EIT). In so doing, it contributes to these countries' modernisation. The emission reductions achieved through JI activities generate a currency called Emission Reduction Units (ERUs). Transfers and acquisi-tions of ERUs have no impact on the stringency of the overall Kyoto objective, as both buying and selling countries have Kyoto commitments. Unlike the CDM, JI is a zero sum game in terms of overall allowed emissions under the Kyoto Protocol. Indeed, any ERU generated through JI is essentially another form of AAU (Streck, 2005b). However, for ERUs to be issued, it must be demonstrated that these emission reductions are additional to any that would occur otherwise.

The rule of *additionality* essentially seeks to demonstrate that emission reductions for which ERUs are issued are only possible in the framework of JI and would not be obtained under a baseline scenario.[35] The demonstration of *additionality* is rather complex in the context of countries like the EITs with rapidly changing economies; and international guidelines do not exist.[36]

JI is to begin in 2008 and includes two tracks affecting the capacity of the participating countries to verify the resulting emission reductions. This capacity is contained in the eligibility requirements stipulated in the Marrakech Accords (see Table 10). However, projects starting as of 2000 can be eligible for JI projects if they meet requirements set in the JI guidelines (to be developed by the JI Supervisory Committee and by national authorities of Annex I countries).

JI Track 2 requires meeting only the first three eligibility requirements indicated in Table 10. It imposes stricter control of JI projects, including a mandatory international verification procedure occurring under the JI Supervisory Committee. This international verification brings the project cycle of JI Track 2 close to that of a CDM project. This procedure could entail higher transaction costs than under JI Track 1 or emissions trading.

JI Track 1 is allowed essentially if the host country has a satisfactory emission inventory and an emission monitoring system is in place. Having this system to adequately account for emissions essentially allows the seller and buyer countries to implement JI rules and modalities without the international oversight of transactions. For example, the additionality of a JI Track 1 project could be determined by the two parties involved according to national guidelines. This could allow more flexibility to use JI Track 1 as an incentive for investments to meet national priorities in relevant sectors.

---

35. The baseline for the JI project shall be established on a project-specific basis and/or using a multi-project emission factor, in a transparent manner and taking into account relevant national and/or sectoral policies and circumstances (COP/MOP1, 2005).
36. We examine this issue further in the section *"Operational" eligibility for JI projects* in this Chapter.

These estimates, developed on a regional basis, exclude the trading that would need to occur within the regional groups here defined. The European Environment Agency's projections of GHG emissions in European countries reveal a significant demand for allowances in some countries (essentially in the original EU-15 countries) and a substantial allowance surplus in others.[37] The IEA (2005a) estimates the potential transfers between 800 and 1 100 MtCO$_2$e annually (including transfers between EU countries).[38]

Governments are expected to play an important role in carbon markets during the Kyoto Protocol's first period in order to offset emission increases from the sectors not covered by existing emissions trading schemes. The purchase of AAUs could be an attractive alternative solution to fill countries' gaps with their Kyoto targets. Indeed, a significant part of international demand could be expressed for AAUs, namely by the governments of Annex I countries. Natsource, a broker and CDM fund operator, estimated that buyer governments would account for between 45% and 73% of all direct international purchases, based on a range of supply and demand scenarios (Natsource, 2003).

The purchase of Kyoto units outside of the existing emissions trading schemes, namely from Russia and Ukraine, both with significant surplus AAUs, will be necessary for compliance of Annex I countries and in particular for the European Union, Japan and Canada. However, whether purchases will be possible and timely will depend on the establishment of attractive and transparent Green Investment Schemes in these countries (see Box 2).

| **Box 2** | Green Investment Scheme (GIS) |

The concept of GIS is not a part of the Kyoto Protocol nor the Marrakech Accords. It is rather a domestic policy option available to countries with surplus AAUs that seek to alleviate concerns of buyers about the environmental integrity of their AAU purchases. There is no internationally-agreed definition of a GIS. The principle of a GIS, however, is rather simple: a GIS would earmark revenues from AAU transfers for environmentally-related purposes (or "green" investments) in seller countries. "The GIS is modelled on JI-type projects but it is more flexible" (World Bank, 2004). Arguably, the exact form, criteria and conditions of a GIS would depend on the needs and preferences of the involved buyers and sellers and may be defined in bilateral negotiations. Thus, the design of a GIS can be made compatible with host-country climate policy priorities. Transfers of AAUs through a GIS transaction could thus be more attractive for host countries than a straight transfer of AAUs (Blyth and Baron, 2003). The GIS could be designed to use a country's surplus AAUs to allow early (pre-2008) and late (post-2012) crediting, and thereby make Kyoto-related activities more feasible and financially viable by providing a carbon revenue stream beyond the five-year commitment period under the Kyoto Protocol.

37. With no new policies, these projections show a trading surplus of some 250 MtCO$_2$e, mainly from EITs, and an excess demand from other countries that would quickly absorb this amount. A more accurate gauge of the trading potential among the 25 EU countries must therefore include the cross-border, inter-regional transactions under the EU Emissions Trading Scheme.
38. These numbers are in line with the estimates by Point Carbon, a company providing information and analysis of carbon markets. Western Europe, Japan and Canada are projected to record a demand of 5.3 GtCO$_2$ over the five-year commitment period, i.e. 1.06 GtCO$_2$ annually (Hasselknippe, 2005).

Russia formally introduced a GIS proposal at the Sixth Conference of the Parties (COP-6) to the UNFCCC in December 2000, albeit with little implementation details. Even prior to Russia's ratification of the Kyoto Protocol, several Russian and foreign experts analysed the potential features and implementation of a GIS in Russia (*e.g.* Tangen *et al.*, 2002; Blyth and Baron, 2003). The various analyses point to the possibility of using the revenues from the sale of AAUs through a GIS for either "hard greening" or "soft greening". These two options are being explored in ongoing discussions on GIS in Russia.

Hard greening: "An AAU is greened when the activities financed through the proceeds of the sale have generated emission reductions measured against […] what would have happened in the absence of the activity" (World Bank, 2004). It is very close to the project-based mechanism of JI Track 1, with the added advantage that there is no fixed time period for emission credits.[39] Activities funded through GIS need not be subject to JI requirements, as they only entail the transfer of AAUs – and not ERUs. However, investors in a GIS will need to be assured that GIS-funded mitigation activities are indeed contributing to environmental protection – their interest in GIS transactions would not exist otherwise. Hard greening corresponds to the so-called "project approach" which could focus on large individual projects, such as capital-intensive energy-efficiency projects or big fuel-switching projects (Tangen *et al.*, 2002).

Soft greening: An AAU is greened when the revenues from its sale go towards "the effective implementation of certain pre-defined activities" (World Bank, 2004), such as, for example, implementing a demand-side management programme, dismantling non-climate-friendly energy subsidies or capacity-building activities related to climate change. Soft greening would be based on the implementation of pre-defined policies or measures that are compatible with – or can facilitate – emission reductions, but it would not be based on the actual emission reductions *per se*. In the Russian case study developed by Tangen *et al.* (2002), the "programme approach" is similar to soft greening. Through this approach, the bundling of a number of small projects can reduce transaction costs.

The above market overview illustrates the interest of many parties to invest in well-identified GHG mitigation projects to fulfil their Kyoto commitments compared to purchases of AAUs – at least at this time. In the case of Russia, where a significant potential for GHG mitigation is available, JI may be an effective tool to encourage such projects. However, JI activities are vulnerable to the timing of Kyoto commitments with crediting starting only in 2008 and official JI institutions under the Kyoto Protocol put in place only since late 2005 at COP/MOP1 (Conference of the Parties to the UNFCCC serving as the Meeting of the Parties to the Kyoto Protocol). The fact that procedures and guidelines are still under development is a major deterrent for project developers today. With limited visibility on purchases beyond 2012, project developers must hurry to allow projects sufficient lead times if they are to expect sufficient returns through the additional revenue stream from sales of emission reductions.

---

39. For JI projects (Track 1 or 2), emission credits are to accrue during the 2008-2012 commitment period.

The notion of "early crediting" was developed by the big institutional and international investors as one solution to these concerns.[40] Countries wishing to attract JI projects prior to 2008 may propose to transfer AAUs for reductions achieved before 2008. This allows JI projects to generate revenues before they can be officially credited for ERUs and thus increase their economic viability. It is nevertheless important that the host country meet the eligibility requirements under Kyoto's emissions trading so that it can assure early investors that AAUs will be transferred for early project-based reductions.

Russian supply via different mechanisms

Given buyers' concerns and reluctance to purchase surplus AAUs, Russia needs to develop a general strategy of supply of AAUs and ERUs in order to maximise the potential advantages from international transactions (Aslanyan and Pluzhnikov, 2003; Laouri *et al.*, 2004; Müller, 2004; Zelinskiy, 2003). How is Russia preparing to do so in practice?

Estimates of Russia's potential for emission reductions supply to the international carbon market remain highly uncertain, over and above the uncertainty related to the level of growth of its emissions between 2008 and 2012. Making an abstraction of "greening" concerns for a moment, Russia's total potential supply of AAUs for the first commitment period of the Kyoto protocol could be directly derived from the projections of its GHG emissions in 2008-12, *i.e.* from 330 to 800 $MtCO_2e$ per year (see Table 8). However, integrating a CO_2 price into economic decision-making in Russia, particularly in the energy sector, via various domestic policies and instruments, could further increase the volume of AAUs available for sale.

The volume of Russian Kyoto unit supply will depend, among other factors, on Russia's expectations of post-Kyoto international climate policy and its possible commitment. Russian officials, however, still need to elaborate their position for the post-Kyoto period. Under the Kyoto Protocol, surplus AAUs may be banked for future use and this provision could be a key component of Russia's strategy for any commitments beyond 2012, given the upward trends in their emissions, energy consumption and economic growth.

The banking of AAUs for use post-2012 under the terms of the Kyoto Protocol would enable Russia to cover its future needs in a rapidly growing economy. It would also provide an opportunity to be better positioned to benefit from possible higher CO_2 prices in the next commitment period.

According to the WEO Reference Scenario, Russian supply could represent over 60% of the total supply in 2010 by EITs, estimated at 1 062 $MtCO_2$ (IEA, 2004b).[41] Russia and Ukraine, without banking, are expected to account for more than 80%

---

40. The "early" JI transactions have been supported by the World Bank since 2002 (PCF, 2002). The "early crediting" was also accepted by the ERUPT tender (De Klerk, 2003). However, this tool is not officially recognised by the JI COP/MOP1 decision on the implementation of Article 6 of the Kyoto Protocol.
41. The Ukrainian potential supply is estimated at 308 $MtCO_2$ by 2010. A larger volume of Ukrainian surplus of AAUs is estimated by other sources, *e.g.* 351-365 $MtCO_2$ (Streck, 2005a) and 388 $MtCO_2e$ (Golub and Marcellino, 2005).

of this supply. Given Russia's potential market power, it could decide to restrict supply to maintain the price of CO_2 at a level that increases Russia's trading revenues on this market. Furthermore, Russia and Ukraine would have an economic interest in seeking to harmonise their AAUs management strategy through restricting overall supply.[42] However, given the political situation that arose over 2005 and into 2006 such a coordinated strategy is unlikely.

Russian officials have expressed their intention of linking each AAU sold from the surplus remaining after banking to a concrete project-based emission reduction activity, in particular via GIS.

Russian authorities are currently discussing possible implementation features of GIS with potential buyers such as Canada, European countries, and Japan. The World Bank is presently working with the Russian government on the terms of reference for a study on GIS implementation in Russia. It was scheduled to begin in early 2006, and could provide useful information and further impetus for this mechanism.[43] However, at the time of publication of this book the study had not yet started due to lack of approval by the Russians.

In the end, the interest of buyers to link purchases of AAUs with environmentally-friendly activities appears to match the preference of Russian government officials to use AAUs to leverage funds for projects, activities and/or measures that may not obtain sufficient budget funding or not be able to attract sufficient investment. For instance, the Russian gas distribution sector, grouping a high number of small-scale opportunities to reduce GHG emissions, appears to be a good candidate to benefit from GIS provided they can be bundled in order to reduce transaction costs (see Chapter 4).

Russian supply will also depend on the role of the Kyoto Protocol flexibility mechanisms in developing Russian national climate policy. Russian officials working on climate policy are currently considering a progressive, "learning by doing" approach, moving from the easiest tools – in terms of implementation requirements – such as JI Track 2 to more sophisticated, yet more efficient, instruments such as JI Track 1, GIS and, eventually, a national ETS. At present, Russia's short term options are limited to JI Track 2, but continued progress in terms of meeting the various eligibility requirements may allow Russia to switch to JI Track 1 before 2008.

The Russian government expects to be in a position to provide the bulk of its Kyoto units supply once it fulfills the eligibility requirements for participation in JI Track 1 and international emissions trading (see Table 10). Full eligibility would also allow Russia to develop an attractive GIS, if buyers remain reluctant to acquire

---

42. The total Russian and Ukrainian revenue from the sale of 50-60% of AAUs could more than double from USD 2.7 to 6.1 billion per year due to the higher price for Kyoto units (Babiker *et al.*, 2002; Haites, 2004). According to the OECD Green Model, market power could increase the price of AAUs by 20% in comparison to the competitive scenario in 2010 (IEA, 2001). However, the model results must be used with care given their stylised representations of the international market for Kyoto units.
43. Some interesting insights for structuring a possible Russian GIS can be found in the proposals elaborated by the World Bank for Bulgaria's GIS (World Bank, 2004).

non-greened AAUs. Such a strategy should ensure the best environmental outcome for Russia, provided the projects in the framework of the JI and GIS generate the GHG emission reductions that are additional to what would otherwise be achieved under a business-as-usual scenario. This will contribute to curbing the national GHG emission trend and maintaining a reserve of surplus AAUs for a future commitment period for Russia.

RUSSIAN CLIMATE POLICY: STRATEGIES AND INSTITUTIONS

Russian climate policy and ongoing progress

The process leading to Russia's ratification of the Kyoto Protocol in 2004 was a long road. The more urgent need to deal with the difficult economic situation of the 1990s largely explains why Russia did not make international climate policy a priority. In addition, Russia's surplus AAUs under the Kyoto Protocol virtually guaranteed compliance (Annex 2 summarises the main stages of development of Russian climate policy).

Brief history of climate policy in Russia

On the other hand, Russia had an interest in maximising the potential economic gains from participating in the Kyoto Protocol flexibility mechanisms. During 1997-2000, the potential Russian (and Ukrainian) revenues resulting from the sale of surplus AAUs were estimated at USD 17-28 billion or 4.6% of Russian and Ukrainian GDP (IEA, 2001). In addition, investments under JI were expected to bring a considerable "multiplier" effect economic and environmental – in a context of capital shortage.[44] Russia hosted several investment projects under the pilot phase of so-called activities implemented jointly (AIJ), begun in 1995.[45] However, the limited success of these AIJ projects served to highlight the institutional challenges facing Russia in using project-based emission reduction mechanisms.

After 2000, with the withdrawal of the United States from the Kyoto Protocol, Russia's interest in ratifying the Protocol decreased given the estimated lower Russian revenues from international emissions trading, as without the United States demand, the price of carbon could drop sharply. In addition, the rapid economic recovery after the 1998 financial crisis revived fears that the legally binding commitments of the Protocol would limit national economic growth. Therefore the development of a national climate strategy and institutional framework for flexibility mechanisms was slow in 2000-04 due to uncertainty regarding Russia's ratification.

At the time of publication of this book, Russia was still in the process of elaborating its climate policy. Currently, the Comprehensive Action Plan to implement the Kyoto Protocol constitutes the guidelines for the development of such a policy and

---

44. According to Golub *et al.* (2004), the efficient use of one dollar of carbon revenue could lead to a multiplier effect equal to attracting 4 USD of investment in the Russian economy.
45. A detailed analysis of the pilot phase is available in Tangen *et al.* (2002) and Korpoo (2005).

corresponding institutional framework. This Action Plan was developed by the Ministry of Economic Development and Trade (MEDT) in mid-2004 just before the ratification of the Kyoto Protocol by Russia, and approved by the government in early 2005. Even though the Action Plan has not been adopted by the government and is not legally binding for implementation, it is an important milestone for the development of Russian climate policy.

Russia's
Comprehensive
Action Plan to
implement Kyoto

The "Comprehensive Action Plan to implement the Kyoto Protocol in the Russian Federation" (Action Plan, 2005) contains two parts: (1) the main measures of GHG emission mitigation, and (2) the institutional provisions necessary to fulfil the obligations under the Kyoto Protocol and the UNFCCC, namely monitoring, reporting and review, as well as meeting other eligibility criteria required to participate in the flexibility mechanisms of the Kyoto Protocol. However, the non-binding nature of the Action Plan weakens its status and creates difficulties for its timely implementation.[46]

Main objectives and measures of GHG mitigation

According to Russian officials, the main objectives of Russia's climate policy are to maximise its contribution to stable energy-efficient long-term economic development, as well as to focus on energy-efficiency measures to reduce GHG emissions in Russia in the medium term.

The Russian Third National Communication (ICCC, 2002) emphasises this link between energy and climate policy. It defined Russia's climate policy as a "set of economically efficient actions which respond to the objectives of energy saving and energy-system upgrading while at the same time lead to considerable reductions of GHG emissions". There are two types of actions: energy-saving policy (technical and organisational measures) both on the producer and consumer side, and the improvement of efficiency in the energy sector itself.[47] This approach, adopted before Russia ratified the Kyoto Protocol, reflects an understanding by Russian policy makers that the success of Russia's climate policy strongly relies on the effective implementation of reforms in the energy sector and substantial improvements in energy efficiency (Silva *et al.*, 2004). Therefore, energy sector reform is critical for Russia's climate policy given the limited success to date of energy-efficiency policies in Russia (see Chapter 1).

The Action Plan uses the emission and energy-efficiency improvement targets of the Energy Strategy of Russia to 2020 and the Federal Programme "Energy-Saving Economy" for 2002-05 and up to 2010.[48] The energy sector targets listed in Table 7 are not binding targets.

---

46. In March 2006, the Russian government asked MEDT and concerned ministries to prepare new proposals for the Action Plan for its approval.
47. This includes replacement and refurbishment of generating capacities to favour more efficient and less energy-intensive technologies, more efficient resource extraction and energy production, as well as modification of the energy mix.
48. Energy Strategy, 2003; Energy-Saving Economy, 2001.

Table 7   Energy sector related actions listed in the Action Plan

	Period	Indicator	Initial source
Reduction of energy intensity of power generation (RAO "UES of Russia")	2004-2006	– 0.08	Energy Strategy
Reduction of associated gas flaring, bcm/toe produced Development of legislative & normative acts introducing mechanisms of gas flaring reduction	2006-2010 1st quarter 2007	10 bcm/toe	"Energy-Saving Economy"
Natural gas savings from gas production to distribution Development of relevant programmes and deadlines	2006-2010 2005 December	47 bcm	"Energy-Saving Economy"
Increase the share of renewable energy in TPES	2004-2010	From 0.1% to 0.22-0.3%	"Energy-Saving Economy"
Development of legislative & normative acts introducing mechanisms of methane emissions reduction in the coal sector	1st quarter 2006		
Refurbishment/replacement of municipal heat networks	2004-2008	From 16% to 30%	

Source: Action Plan, 2005.

The draft Federal Programme "Energy-Saving Economy" for 2006-10, which is currently under review, establishes new efficiency targets and integrates JI and GIS mechanisms as new instruments of the programme's implementation. At the time of writing, the draft version of this programme was not publicly available.[49]

The deadlines set in the Action Plan indicate the sectors considered by the government as priority areas for implementing GHG emission reductions measures.[50] These priorities reflect the government's aim to target investments in flexibility mechanisms in sectors that would have difficulty in attracting investments under normal business conditions. For example, investments are needed in energy-efficiency improvements in the residential sector, district heating, coal bed methane, etc. where major benefits can be achieved at relatively low cost. Similarly, one could expect that the Russian government would seek to use flexibility mechanisms in the gas distribution sector, which has a limited capacity to attract finances with many energy-efficiency investment barriers to overcome (see Chapters 1 and 4).

Institutional provisions in the Action Plan

The value of the Action Plan largely lies in the deadlines the government has set for itself in preparing to meet the flexibility mechanisms' eligibility requirements and in establishing a national framework. Equally important is the indication of the responsibilities of Ministries and Agencies for achieving these objectives. However, as of early 2006, the Russian institutional framework for climate policy had considerable lacunae mainly due to the ongoing bargaining among Russian authorities over the responsibility and control of climate policy activities in Russia.

---

49. This new programme seeks to provide new economic incentives for energy savings taking into account the slow implementation of previous programmes.
50. The deadline set for measures to "reduce the burning of associated gas" is the first quarter of 2007, is much later than the deadline of December 2005 for "measures to reduce gas losses in the gas industry", and the first quarter of 2006 for "methane emissions in the coal industry".

Table 8 shows that only one (although essential) eligibility requirement was fulfilled by Russia, namely the ratification of the Kyoto Protocol. The tight timeline set to meet the eligibility requirements in the Action Plan (end 2006) indicates the MEDT's understanding of the need to rapidly fulfil these eligibility requirements in order to have access to the JI Track 1 procedure (see Box 1) and AAUs transactions. Meeting this timeline may allow Russia to submit its compliance report to the Compliance Committee of the UNFCCC by the end of 2006, in line with other Annex I countries, and to acquire eligibility status in early 2008.[51]

Russia's ability to meet this ambitious timeline has been put into doubt by its inability to meet similar timelines set last year. At its first two meetings (July and November 2005), Russia's Inter-Agency Commission on the Implementation of the Kyoto Protocol agreed to extend the deadlines it had set earlier.[52] However, in early March 2006, the government decided to set new deadlines for two key technical tasks: establishing its national registry (June 2006) and a national system for estimating GHG emissions (July 2006). The Inter-Agency Commission must also submit estimates of its financial needs for governmental approval in order to ensure the further development and functioning of these tools.

Table 8   Russian deadlines to fulfil eligibility requirements for participation in the Kyoto Protocol flexibility mechanisms and in alternative schemes

	Eligibility requirement	ET (& GIS)	JI T-1	JI T-2	Status	Timeline
1	Party of the Kyoto Protocol	•	•	•	•	
2	Assigned amount calculated and recorded	•	•	•	no	2nd Q 2005
3	National registry for assigned amount (Kyoto units transactions)	•	•	•	no	3rd Q 2005 (new: 2nd Q 2006)
4	National system for the estimation of annual GHG emissions/ removals	•	•		no	2nd Q 2006
5&6	National annual report on the assigned amount and on the latest GHG inventory*	•	•		no	3rd Q 2006
	Specific issues for JI:		•		no	2nd Q 2005
7	Designated focal point for JI project approval					(new: 2nd Q 2006)
8	National guidelines and procedures for JI project approval		•		no	2nd Q 2005

* Including additional information on assigned amounts and any adjustments to assigned amounts
Sources: based on Marrakech Accords, 2001; Action Plan, 2005; Russian Government, 2006a, b.

---

51. The Compliance Committee (Enforcement Branch) of the UNFCCC determines if the requirements are met on the basis of the Initial Report on Compliance submitted by the country. Automatic eligibility is attributed 16 months after submission of the Initial Report (unless the Committee determines a requirement is not met) or earlier if the Committee decides not to proceed with any questions.
52. According to an official of the Inter-Agency Commission, deadlines were pushed back, in particular for the national GHG inventory.

In order to use flexibility mechanisms or GIS, the Russian Duma will need to define legislation to provide the government with the authority and responsibility for managing AAUs at both federal and regional government levels, if such a choice is made.[53]

Defining the legal basis of any operations with Kyoto units involving Russian private entities is important. Under Article 3 of the Kyoto Protocol, allowances are assigned to governments by virtue of international law. The government can translate its target and limitations through the allocation of allowances to private and public entities, using national climate policy instruments (Streck, 2005b). Thus, the legislator would need to determine the principles of transfer of rights to manage AAUs to private and public entities and the entities that are eligible for flexibility mechanisms.

The rights to perform AAU and ERU transactions can be established either through legislation or through government decisions, authorising an entity to participate in a project and to hold and transfer ERUs. The MEDT was to submit to the government its proposals for the necessary amendments to Russian law to create a legal basis for the JI mechanism by mid-May 2006. The draft MEDT proposals for the JI procedure were based on the Russian law on Capital Investment and on related technical regulations.

The Action Plan does not specify any instruments of national regulation of GHG emissions or include provisions for developing AAUs allocation rules to public or private entities - at least until 2008.[54] It reflects the current debate on priorities and choice of instruments described below. The climate policy under consideration contains a set of specific instruments of GHG emissions control and mitigation that could create new incentives to reduce emissions in the energy sector through greater energy efficiency. The climate-policy instruments include, for example, regulatory instruments (carbon tax, norms, and standards) or market-based instruments (domestic emissions trading, Kyoto Protocol flexibility mechanisms).

National climate policy and choice of instruments still uncertain

Russia's ability to capture synergies between climate and energy policies will depend on both its climate strategy and on the choice of instruments. There is an ongoing debate within the government, between proponents of an administrative regulation of greenhouse gases (*i.e.* emission fees) and proponents of market-based instruments, including a domestic emissions trading system similar to the European Union and considered by a number of other countries.[55]

---

53. The Russian Federation includes 89 subjects or regions.
54. The instruments of the implementation of the Action Plan are similar to those listed in the Third National Communication, namely the gradual reduction or elimination of market disproportions and the use of market instruments stimulating reforms in the energy sector.
55. The European Union Emissions Trading Scheme (EU ETS) introduced in January 2005, is currently the biggest emissions trading system encompassing approximately 12 000 plants across the EU-25 and about 45% of the EU total CO_2 emissions. The EU ETS includes only CO_2 emissions considering the complexities of measurement and verification for other GHG, in particular for CH_4. IEA (2005a) provides a detailed analysis of EU ETS and other emissions trading schemes.

Regulating greenhouse gas emissions

The current practice of Russian environmental policy favours the administrative regulation of pollutants (see Box 3). Incorporating GHG emissions in the existing list of regulated pollutants would mean relying on the existing monitoring and control system under the responsibility of Rosteknadzor - the federal agency in charge of environmental, technological and nuclear monitoring and supervision. This option would not require existing legislation to be amended, but it would be necessary if a domestic emissions trading system was established.

In mid-2005, Rosteknadzor and MEDT agreed to increase the existing fees for CH_4 emissions by a factor of 1 000, albeit from an insignificantly low base – thereby providing an incentive to reduce CH_4 emissions by industry.[56] Namely, these fees cover CH_4 emissions from leaks at equipment and components along the natural gas system and also CH_4 contained in the associated gas flared by oil companies.[57] The fee level for CH_4 emissions below the permissible emission limits (see Box 3) is set at 50 roubles per tonne while the emissions above these limits are charged at a rate of 250 roubles per tonne. The effectiveness of such an approach will hinge largely on whether and how monitoring agencies will be able to perform their function.

According to IEA discussions with Russian climate policy makers and as reflected in the Action Plan, the choice of climate policy instruments depends largely on the government's evaluation of the financial and technical capacities of different industries to implement GHG emission reductions measures, as well as the potential for these measures to contribute to the sustainable development goals of Russia. In mid-2006, the thinking within government, and especially within the MEDT, appears to be that the Russian oil patch needs little incentive to attract outside investment compared to other sectors.[58] With the same logic, we could assume that the gas transmission sector, which benefits from growing export prices, also does not require additional financial incentives to implement energy-saving measures. Indeed, pollutant fees could be increasingly used to regulate emissions in these sectors.

The IEA questions whether an increase in pollutant fees provides sufficient incentive for oil and gas companies to increase their energy-efficiency investments. This is all the more uncertain given the limited enforceability of such measures in the past. As discussed in Chapter 1, the investment decisions of Russian oil companies, independent gas producers and Gazprom encompass a wide range of factors, such as increasing their export potential in the context of low domestic gas prices and dealing with the structural rigidities of the gas sector including the limited competition in the upstream. While fees are coherent with the widely-accepted "polluter-pay principle", limited effectiveness in Russia suggests the need for complementary measures.

---

56. Russian Government, 2005.
57. This does not exclude the use of other "flexible" options to reduce methane emissions in Russian industries.
58. Discussions with MEDT officials in July 2005.

In the short term, project-based flexibility mechanisms, such as JI and GIS, could provide necessary incentives for energy-efficiency gains via GHG emission reduction revenues.[59] This would also match the interests of Annex I investors/buyers to purchase ERUs in Russia at an attractive price. When institutional capacities become available, the use of a domestic emissions trading system could ensure a more efficient distribution of the GHG emission reduction effort among participants via a carbon price.

Box 3   The system of fees for pollutant emissions in Russian environmental policy

A comprehensive system of environmental quality standards forms the basis for granting permits and setting fees at the federal level. The most important standards are the Maximum Allowable Concentrations (MACs), which establish maximum values for peak and average concentrations of environmental pollutants. The MACs cover over 500 standards for air pollutants, over 2 500 for water pollutants and over 100 for soil pollutants. They are often based on public health requirements or public safety criteria developed by the World Health Organisation (WHO) and they thus tend to be very stringent. The government sets Maximum Permissible Emissions (MPEs) based on MACs for enterprises, municipal treatment facilities and other stationary sources of pollution. Standards have also been set for concentrations of harmful substances in emissions from mobile sources. To reflect current technical and economic limitations, less strict Temporarily Permitted Concentrations (TPCs) and Emissions (TPEs) are used as an intermediate step in meeting the stricter MPEs.

The main economic instrument of environmental policy is the imposition of fees for pollutant emissions and discharges. All polluting sources are subject to a base fee proportional to their emissions or discharges. Multipliers or "ecological coefficients" raise the per-unit charges under specific conditions, designated as environmental emergencies or disaster zones. When emissions exceed the MPEs but are below the TPEs, the base charge is multiplied by five; when they exceed TPEs, the multiplier is 25.[60]

The fee system, which was very effective in the early 1990s, lost much of its effectiveness due to rapid inflation. In principle, indexing emission charges should ensure that it is more expensive to pollute than to comply. Although the fees were revised regularly over the 1990s, they failed to keep pace with inflation. Between 1990 and 1996, the real worth of pollution charges decreased by a factor of 20. For some of the more prosperous companies, especially oil refineries, pollution fees were so low as to be insignificant; for others, like uneconomic coal mines that sustained heavy losses, the charges simply remained unpaid.

---

59. Revenues and profits from project activities also lead to additional tax revenues, potentially easier to track and collect than pollution fees.
60. With the approval of federal and regional environmental authorities, enterprises may offset their emission charges against all or part of the value of environmental improvements made at their own expense. Offsets in some regions may exceed cash transfers, and they can be important even in comparatively well-off regions.

Domestic emissions trading for Russia

Another possibility for Russian climate policy would be to opt for incentive-based options to implement GHG mitigation objectives, such as the establishment of a domestic emissions trading scheme (ETS). This option is supported by the MEDT and also by environmental organisations (RREC, 2003; WWF, 2005) as well as by large industrial players in Russia, via the National Carbon Union (NCU).[61] In the Programme of Social and Economic Development of Russia for the Medium Term (2006-08), approved by the government in January 2006, the MEDT included the principle of emission trading as a possible tool to control environmental pollutants, including GHG (MEDT, 2006b).

MEDT supports the idea of the development of a Russian "cap-and-trade" ETS as one of the possible options for domestic trading, with caps to be agreed through voluntary agreements between Russian companies and the government. Companies could sell GHG emission allowances if they reduce emissions below their voluntary targets. These targets could be relative, *i.e.* expressed as tonnes of CO_2 per unit of output. Such relative targets provide more flexibility and lower economic risk for participating companies (Blyth and Bosi, 2004), as unexpected growth in output and emissions imposes less cost than if emission caps were fixed at a given absolute level. As for other existing schemes, at early stages a Russian ETS could cover a limited number of sectors with large stationary sources, for example the electricity[62] and gas transmission sectors, and then progressively extend coverage to other sectors.[63]

The MEDT is examining the option of basing voluntary targets for possible ETS participants on technological standards for new equipment (Gavrilov, 2005b) with the objective of stimulating the turnover of energy-intensive equipment stock, especially in the energy sector. This approach would avoid the premature retirement of long-lasting energy equipment before the end of its technical life. In Russia, a high percentage of equipment is currently being used beyond its technical lifespan and would need to be replaced anyway. Thus, in order to become an efficient instrument of climate policy, in terms of environmental performance, an ETS must state sufficiently stringent objectives, at a minimum below the business-as-usual level of GHG emissions of the participating sectors.

It is worth noting that several Russian experts have expressed the view that the creation and management of a domestic ETS would probably be easier and cheaper than the implementation of the bureaucratic mechanism of selection, approval, verification and monitoring of individual projects required under JI. Many more would likely agree with this once the regulator determines the design and allocates the emission permits among participants of ETS (Tietenberg, 1997; Kolstad, 2000; Stavins, 2000; Ellerman *et al.*, 2003; IEA, 2001). However, getting there is not straightforward. Practical experience gained during the preparation phase of the

---

61. NCU regroups the sources of more than 30% of Russian GHG emissions (see www.natcarbon.ru).
62. The Russian electricity sector represents nearly one third of national CO_2 emissions and has developed a considerable capacity in preparing Kyoto-related projects via its Energy Carbon Fund established in 2000.
63. See IEA (2005a) for a detailed discussion on the potential to enlarge the ETS beyond large industrial sources of emissions (i.e. transport sector and civil aviation) and on related difficulties.

EU ETS shows that implementing such a scheme is not simple and the initial "effort" to set-up an emissions trading scheme can be significant and time consuming. These efforts include:

■ The definition of the system's coverage and boundaries (which sectors should be included, what thresholds should be set to have a manageable number of sources in such a scheme).

■ The collection of data necessary to define emission objectives, the definition of their stringency, and principles for allocating emission allowances.

■ The establishment of a reliable monitoring, reporting and verification system in order to assess the compliance of participants to their emission targets.

Russia's extensive industry, geography and administrative structure provide significant challenges for rapid implementation of a trading system resembling that of the EU. Currently, no detailed study is available on a specific design of a Russian domestic ETS.

On the other hand, Russia could learn from the EU ETS experience. This could help Russia to move progressively towards a system that would be at least technically compatible with other emerging domestic GHG trading systems and facilitate the linking of the systems at a later stage if necessary.[64]

Linkages with other trading systems are important to Russian industry in order to support a national ETS. The possibility of direct participation in international emissions trading would avoid inherent transaction costs associated to the project cycle of JI (or a project-specific approach of a GIS). The political recognition of emission allowances implemented in Russia from any ETS is important for linking a Russian ETS to other ETSs. Confidence in the Russian allowance would presumably depend on the ability of the Russian authority responsible for ETS to enforce the monitoring, reporting and verification procedures. Concerns may also arise due to different possible sectoral coverage of a Russian trading scheme, and the stringency of the emission reduction targets.

Some Russian experts estimate that a national ETS could start by 2010. More optimistic experts suggest that it could take as little as 2 years from the moment the political decision is taken (Mielke *et al.*, 2004). However, in addition to data collection and political discussion on acceptable allocation levels – much time is needed to prepare the legal framework.

At present, Russia's environmental law does not include tradable emission allowances and some Russian legal experts consider that it would be necessary to make several amendments to allow this. The EU ETS experience suggests that introducing an emissions trading scheme could raise the following legal issues (Streck, 2005a):

---

64. For more details on linking issues see Blyth and Bosi (2004), IEA (2005a).

■ The system should be constitutional.

■ Its relationship with other environmental policies in Russian law should be clarified.

■ Allowances should be defined: would they be a property right, a subsidy, a financial instrument?

■ How should allowances be valued in companies' balance sheets and treated fiscally?

These issues are not specific to Russia, and experience from other countries shows the complexity of adapting new notions to the domestic legislative system – such as AAUs, transferable emission allowances, ERU, etc.[65]

Implementing a national ETS would also require legally binding rules and obligations to monitor, report and verify GHG emissions. This legislation could be difficult to enforce in Russia given the dominant position of energy monopolies in Russia which are key GHG emitters. Progress on regulatory reforms in Russia's energy sector could provide the necessary impetus to move forward on establishing an ETS.

The Russian government also relies on foreign expertise to develop its institutional climate policy framework. For example, the implementation of the EU Commission's TACIS programme which aims to assist Russia in preparing its national GHG inventory and in developing a framework for JI projects began in September 2005. This programme could improve Russia's prospects of fulfilling the eligibility requirements before the beginning of the Kyoto Protocol's first commitment period. However, to meet these requirements on time depends on the priority of the Kyoto Protocol in Russia's political agenda and on continued support from the Russian government.

PARTICIPATION IN FLEXIBILITY MECHANISMS: RUSSIA'S CAPACITY

"Technical" eligibility requirements

The development of reliable GHG inventories is currently one of the more complex, but affordable issues for Russian eligibility. In the Third National Communication (2002), Russia has provided estimates for its GHG emissions in 1990-99, only partly based on IPCC guidelines for National Greenhouse Gas Inventories (IPCC, 1997). The Second UNFCCC In-depth Review (Silva *et al.*, 2004) concluded that the main "implementation" issues were related to insufficient funding. Historically, the government has paid limited attention to this topic due to lack of interest in management of AAUs (Yulkin, 2005).

---

65. According to Streck (2005a) this was a common issue in all EU countries participating in the EU ETS.

Development of
the national GHG
inventory and
definition of the
assigned amount

According to CENEf and PNNL (2004), which compiled an unofficial GHG inventory for Russia, existing information should be sufficient to elaborate "accurate and complete inventories of CO_2 emissions", especially from fuel combustion. In addition, the 2004 Report on the In-depth Review indicated that Russian experts had already begun work on preparing a sectoral approach for the energy sector (Silva *et al.*, 2004). The main "technical" problem of Russia's inventory is in fugitive emissions from fuels, which are one of the "large key sources of emissions" in Russia (representing more than 7% of the total), including estimates with a high level of uncertainty.[66]

In the oil and gas sector, two methods can be used to calculate fugitive CH_4 emissions:

■ The Tier 3 method – the rigorous source-specific evaluation, requiring detailed inventories of infrastructure, and detailed bottom-up emission factors.

■ The Tier 1 method using aggregate production-based emission factors and national production data. This approach is susceptible to substantial uncertainties and should be used as a last resort option.[67]

For both methods, industry should be disaggregated into its segments and sub-categories so that emissions can be evaluated separately for each of these parts. The gas industry segments include: wells, gas production, gas processing, gas transmission, storage and gas distribution.

Given that fugitive emissions from oil and gas systems are one of the key sources of emissions in Russia, the Tier 3 method should be implemented. Indeed, additional efforts are necessary to develop an inventory of infrastructure and facilities and to determine specific non-CO_2 emission factors. The application of rigorous bottom-up approaches may be difficult and costly. It is useful to point out that Kyoto-related activities, such as those under the framework of a GIS, could contribute to improving the quality of the GHG inventories in areas of greatest uncertainty.

The first priority is to determine the appropriate country and source-specific emission factors. The uncertainty level associated with CH_4 emission factors is not specific to Russia's natural gas sector. However, given the large scale of Russian natural gas systems and the insufficient number of independent on-field measurements, available CH_4 emission data is not representative of the entire system. This was demonstrated in the gas transmission sub-sector by several measurement programmes implemented since 1996 (see Chapter 3). In the gas distribution sub-sector, which has received only limited attention in the past, only partial measurement programmes have been conducted (see Chapter 4).

---

66. The Third National Communication (ICCC, 2002) estimates CH_4 emissions from both the oil and gas sectors at just below 9% of national GHG emissions in 1999. According to our estimates, in 2004, the gas sector alone accounted for the same share of total GHG emissions.
67. According to Russian experts, the use of the refined Tier 1 emission factors based on North American data may result in high uncertainty given the differences in the design, operating practices and state of the oil and gas industries. Nevertheless, these are the best comparable options, and are systematically used in this study.

The second priority for a good quality inventory is to collect activity data by source of fugitive emissions. For CH_4 emissions in the natural gas sector, for example, it concerns the number of leaking components from compressor stations, transmission pipelines or the number and age of metering and regulating stations in the gas distribution network. We show in Chapters 3 and 4 that it is difficult to obtain detailed representative activity data from both the transmission and distribution sub-sectors.

In Russia's gas distribution sub-sector, information on existing equipment stock is incomplete. The high level of commercial losses and the lack of metering equipment imply significant uncertainties in estimates of fugitive emissions. However, as demonstrated in Chapter 4, Russian gas distribution companies have indicated an interest in developing good quality measurement methodologies necessary to monetise the supplemental Kyoto-related benefits through efficiency projects.

Involving oil and gas companies in developing sectoral GHG inventories is useful in implementing the Tier 3 approach, given the set of minimum required activity data (IPCC, 2000). For the gas transmission sector, for example, taking into account the age of facilities and their maintenance records, this data includes:

■ For equipment leaks – facility/installation counts by type, process used at each facility, equipment component schedules by type of process unit, and gas/vapour compositions.

■ For gas-operated devices (*e.g.* compressors) – schedule of gas-operated devices by type of process unit, the gas consumption factors, etc.

■ For pipeline leaks – the type of piping material, length of pipeline.

Currently, both Gazprom and RAO UES of Russia are continuing their inventory-related work with regard to their own systems, equipment and operations, including work on related emission factors. Table 4 shows the aggregate inventory of Gazprom's GHG emissions in 2000. Gazprom officials have indicated that several other elements of the company's GHG inventory would be made public only when the Russian government establishes clear and predictable objectives and rules for the country's climate policy.[68]

The progressive refinement of emission inventories for the oil and gas sectors will provide the possibility of capturing more accurately the impact of specific GHG control measures. The improvement of a GHG inventory in these sectors could also significantly enhance their opportunities to take advantages of the flexibility mechanisms. An internationally accepted methodology of methane emission estimates – which has yet to be developed – could facilitate the implementation of Kyoto-related projects in Russia's natural gas sector.

---

68. Discussion with Gazprom representatives in July 2005.

The progress made in developing GHG inventories in some Russian regions and by large private and public companies provides an additional possibility to crosscheck the national inventory, especially for the major emitting regions (Silva *et al.*, 2004).[69]

Experts generally agree that the combined capacities at regional and federal levels are sufficient to develop a reliable national inventory for Russia. For the 1990 emissions inventory – when the Russian Federation was not an independent state – the accounting of energy consumption in the USSR was used which was generally of good quality. In fact, the estimated level of energy-related CO_2 emissions in 1990 is considered accurate, but the main problem is the disaggregation of the total amounts of USSR's emissions to extract the estimates of non-CO_2 emissions. Determining Russia's assigned amount will be based on completion of the 1990 inventory.

At the time of publication of this book, Russia had not yet submitted its national inventory report or its common reporting format (GHG emission data series) to the UNFCCC, having missed the annual 15 April deadline for 2006. It would be a positive step if Russia were to submit its first annual inventory report later this year, as it is the only Annex I country that has not done so since 2004.[70]

Development of the GHG registry

The registry is a national log of a country's assigned amount and other Kyoto units, and transfers and acquisitions of Kyoto emission units (*e.g.* AAUs and ERUs). It is compared to the national inventory (which keeps track of actual GHG emissions) in order to establish compliance. For Russia to be in compliance, for example, the amount of Kyoto units in the registry would need to be equal or greater to the actual emissions as in the national inventory.

The establishment of the registry is a technical issue, and no major problems are anticipated for Russia. According to experts, the Ministry of Natural Resources, responsible for the implementation of the registry, has the necessary funding to ensure the adaptation and functioning of the software. In mid-2005/2006, Russian experts were considering other proposals by existing foreign registries that could be adapted to meet Russia's conditions with the assistance of the European Commission's TACIS project.[71]

"Operational" eligibility for JI projects

It is generally agreed by Russian officials that although more bureaucratically burdensome, JI Track 2 is the short-term answer for GHG emission reduction activities in Russia. They will focus on this for the next 6-12 months while implementing the various tasks set out in their Kyoto Action Plan.

Currently facing JI Track 2

What are the required institutions and procedures to meet JI Track 2 requirements? This procedure was clarified by the eleventh Conference of the Parties/ first Meeting of the Parties to the Kyoto Protocol (COP/MOP 1) in Montreal at the end of 2005. The MOP 1 approved the Marrakech Accords, which determine the broad principles and guidelines for flexible mechanisms, including the JI Track 2 project-cycle.

---

69. CENEf has conducted or planned ten regional inventories in Archangelsk, Novgorod, Chelyabinsk, Sakhalin and Republic of Hakassia and others (Leneva, 2002). These regions represent over 13% of Russian industrial volume and 12% of the population in 2001 (Korpoo, 2004).
70. See UNFCCC Web site on National Inventory Submissions.
71. As pointed out in the section on institutional provisions of the Action Plan, the government requested the establishment of the national registry by 1 June 2006.

Under JI Track 2, the projects are implemented according to the international JI guidelines based on the COP/MOP decisions and on relevant CDM experience. The Accredited Independent Entities (AIEs) accredited by the Joint Implementation Supervisory Committee (JISC) are responsible for the determination and verification of GHG emission reductions (ERUs) by JI Track 2 projects.[72]

If eligible for JI Track 2 (see Table 8) at the operational level, the host country has to establish the National Designated Authority (DNA) responsible for approving JI Track 2 projects. According to the government decision in mid-March 2006, the Russian DNA should be nominated by mid-May 2006. At the same time, legislative amendments should be ready to integrate JI procedures into Russian law. Although a draft version of the national JI rules is not yet available for public discussion, MEDT officials have stated their preference for "simple, transparent procedures closely resembling a mere registration process" for JI projects (Gavrilov, 2005a).

At the time of publication of this book, the international JI guidelines were still under development by the JISC. As for CDM projects, JI project design should include an appropriate baseline and monitoring plan, as well as demonstrate the additionality of emission reductions. According to COP/MOP1 decisions, investors in JI projects could also use the methodology already approved by the CDM Executive Board. The CDM additionality tool example (see Box 4) could provide some indications on possible future rules for Track 2.

There is also an interest in developing more streamlined JI procedures to avoid difficulties and delays already experienced in implementing CDM procedures, but taking account of the different nature of JI and CDM.[73] The JI projects function under the cap of the Kyoto commitments of Annex I countries. Whereas the transfer/acquisition of the ERUs related to JI projects is a "zero sum game", CDM projects add supplemental (new) Kyoto units to the total amount of countries' assigned amounts.

The development of streamlined JI procedures/rules is important given the experience to date of CDM project activities, suggesting that the JI Track 2 project-cycle could take over 10 months to complete (estimates of IETA, 2006). Standardisation and development of specific rules for small-scale activities are possible ways of streamlining the JI Track 2 project-cycle and reducing transaction costs.[74]

Standardisation relates to using multi-project emission factors to establish project baselines and/or developing similar procedures or formats to determine ERUs and the development of the project design document. This work, in particular on multi-project emission factors at the sectoral level, should be implemented in co-ordination with the authorities of host counties responsible for developing the national JI guidelines. This is discussed below when considering possible future guidelines for JI Track 1.

---

72. Projects could also be reviewed by JISC if requested by participants of host countries.
73. There is less of a difference with JI Track 2 as a host country is not fully eligible and does not have all the necessary tools to control its compliance.
74. The transaction costs specific to the project-based flexibility mechanisms could be related, for example, to the cost and time-consuming procedures of approval, registration, monitoring and issuance of ERUs, as well as the cost of verification and validation of specific methodological issues (baseline study, demonstration of additionality, monitoring plan, etc.).

The JISC has been asked to develop guidelines for small-scale JI projects. This is a new concept for JI and several important issues will require further analysis in the context of JI host countries, *i.e.* the clarification of small-scale JI characteristics, the appropriate scale of projects and/or bundled activities (for example, the relevance of increasing the scale for energy-efficiency projects, of "unlimited" bundling, etc.), the need for specific rules and capacities in the host countries to support small-scale projects.[75] According to the JISC work plan, the first decisions on the relevance of small-scale rules for JI activities could appear by the end of 2006 (after the fifth JISC meeting). The need for elaboration of these rules was raised by Russia during the COP/MOP1. The streamlining procedure for small-scale JI projects could become an attractive framework for GHG emission reductions in sectors with a high number of dispersed emission sources, such as the residential sector, district heating systems, renewables and gas distribution networks (see Chapter 4).

| **Box 4** | Additionality demonstration for CDM projects |

Under the CDM, the determination of the emission baseline and of the *additionality* of proposed project activities is assessed through two distinct but linked procedures. The "Tool for the demonstration and assessment of additionality" has been developed under the CDM Executive Board (CDM EB, 2004). It provides a step-wise approach that project proponents can use to demonstrate and assess the additionality of their proposed CDM project activity, *i.e.* that emissions are reduced below those that would have occurred in the absence of the registered project activity. Should the procedures for a JI project follow a similar route, steps 1 through 4 of the CDM additionality tool (see Figure 14) could provide useful material for developing possible future JI additionality rules, particularly for JI Track 2 projects.

Step 1: The alternatives (baseline) of the project activity shall be consistent with all applicable legal or regulatory requirements. However, if these requirements are systematically not enforced, this may be taken into consideration.

Step 2: The CDM project activity, without the carbon revenues, should be less economically or financially attractive than the baseline scenario activities. Instruments: simple cost or investment comparison analysis (internal rate of return, net present value indicators) or financial benchmark analysis.

Step 3: The barrier analysis demonstrates the barriers that would prevent the implementation of the CDM project even if it is economically attractive, such as investment barriers (access to funding, absence of financial incentives), technological barriers, and barriers due to prevailing practices (first-of-its-kind activity).

Step 4: The analysis of the common practice demonstrates the extent of the diffusion of the proposed activity (technology or practice) in the relevant sector and region.

---

75. Extensive information about small-scale CDM projects including definitions, methodologies and examples of approved projects is available on the CDM Web site.

Figure 14   Additionality tool of CDM projects

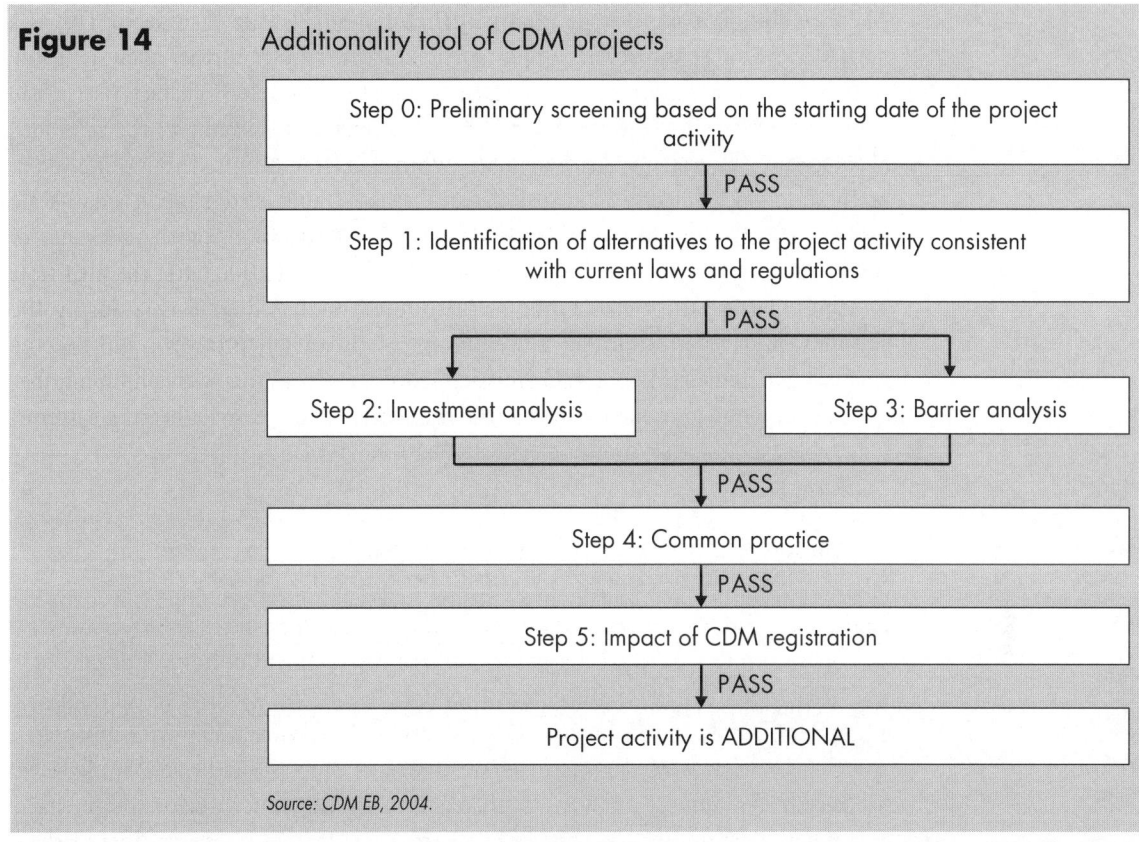

Source: CDM EB, 2004.

Russian authorities seek to improve the competitiveness of the Russian JI framework in comparison with other JI host countries, namely by gaining access to JI Track 1.[76] In order to achieve this, Russia must establish several institutions and procedures in order to demonstrate its capacity to estimate and monitor GHG reductions of JI projects.

Moving toward the JI Track 1

Under the JI Track 1 procedure, national authorities of the host country and the investor/buyer country would have more decision making power than under JI Track 2. According to the Marrakech Accords (2001), the host country eligible for JI Track 1 may verify and monitor the GHG emission reductions by JI projects and issue the appropriate quantity of ERUs.

The approval of the JI projects as well as the verification and reporting would be under the responsibility of the DNA, acting as a national focal point. However, during 2005, even before the establishment of the DNA in Russia, MEDT issued letters of endorsement for 30 JI project proposals. The periodic verification of the ERUs generated by JI projects can be transferred to AIEs accredited according to the standards and procedures of the Marrakech Accords (2001), namely those acting in the framework of JI Track 2.

---

76. In February 2006, Point Carbon ranked Russia tenth among ten JI host countries. This rating reflects: national investment climate, climate policy and institutional JI framework, current JI project experience and project proposals.

Methodological issues: national guidelines for baselines and additionality

Under JI Track 1, the host country may either develop its own national guidelines for JI activities or use the international guidelines developed by JISC. The international JI Track 2 guidelines could be considered possible common ground by host countries to ensure the transparency (comparability) of projects and activities.[77]

In co-operation with industry, which has the necessary sectoral technological and economic knowledge, the DNA could undertake the preparatory work to make the national JI procedure more attractive to the investor:

- Firstly, the preparatory work could standardise JI methodology, namely project baselines.

- Secondly develop a national tool to determine additionality (see Box 4), which would take into account the national and/or sectoral specificity (for example, consideration of national and/or sectoral policies and circumstances in baseline scenarios).[78]

The methodology should be both accurate and streamlined in order to minimise JI-related transaction costs.

At the sectoral level, many analytical efforts have been made by international bodies (*e.g.* see extensive IEA/OECD work in this field), various carbon funds and tenders (*e.g.* World Bank Prototype Carbon Fund, Netherlands' ERUPT/CERUPT, BASREC, etc.) and project developers. In the gas sector, Hanle (2003) provides the first detailed analysis of baseline development – including Russian examples – as well as discussing issues of additionality.[79]

The standardisation of baselines (*e.g.* in terms of specific emission factors in tCO_2 per bcm) for the gas sector may be rather limited due to the site-specific factors which determine the business-as-usual scenario. However, there may be more scope for standardising baseline methodologies. For example, Hanle (2003) provides interesting proposals on the decision tree structure for project developers for three types of projects: fugitive emission reductions, flaring reduction and efficiency improvement projects (especially of compressors). For each type of project, Hanle (2003) determines a minimum requirement for data, the possibility of using default emission factors and/or methodology given the different treatment of greenfield and brownfield projects. In our study we develop and complement Hanle's approach by focussing on the technical and economic aspects of gas transmission and distribution and gas flaring in Russia.

---

77. These guidelines are also viewed by some investors/buyers as a "risk management tool" for JI Track 1 projects if the host country cannot ensure its eligibility over the crediting period (for example, it may not meet the eligibility requirements for JI Track 1 on time and thereby put the JI activities at risk).
78. In the CDM framework it was clarified by the Executive Board Report 22 (CDM EB, 2005a).
79. Other Russian case studies in the oil and gas sector are available in GGFR (2003), BASREC (2003).

Other examples of baseline methodologies are also available from the list of approved CDM methodologies:

- The baseline methodology AM0009 for the Rang Dong project of "Recovery and utilisation of the associated gas from oil wells that would otherwise be flared" in Vietnam (CDM EB, 2005c).

- The baseline methodology AM0023 for the "Leak reduction from natural gas pipeline compressor or gate stations" in Moldova (CDM EB, 2005b).

Demonstrating additionality could be the main methodological challenge to using of project-based mechanisms in Russia's natural gas sector. Efficiency projects in the energy sector are arguably the only sustainable business-as-usual development scenario for this sector and this is also stated as a priority in Russia's national energy strategy. With the progressive increase in energy prices, efficiency projects in the gas sector should become more and more economically attractive on their own due to gas savings, regardless of the substantial benefits of GHG emission reductions. Under such a scenario, traditional financial analysis may not be appropriate, making the demonstration of additionality of efficiency projects particularly complex and challenging.

However, the demonstration of additionality in itself is somewhat controversial, as highlighted by the paradox stated by Grubb: "The difficulty of ensuring that crediting reflects real and additional emission reductions is compounded by the paradoxes that most 'cost-effective' projects may be the least 'additional' and that strict project additionality would give perverse policy incentives" (Grubb *et al.*, 1999). It can be argued that a project should be financially solid to ensure long-lasting GHG reduction benefits – the objective of project-based mechanisms. However, in practice so far, the weaker the financial aspect of a project activity, the greater its chances of being considered additional. A too stringent additionality demonstration can eliminate a share of eventual projects and give a disincentive to Kyoto-related investment. However, the use of too compliant additionality criteria would result in the crediting of business-as-usual projects and in the misuse of climate-related resources. In the case of Russia's gas sector, however, it is important to understand that energy-saving and efficiency projects are often not implemented due to institutional or market barriers. We discuss in greater detail these important factors, which could potentially be used to demonstrate project additionality, in the sectoral chapters of this book.

JI prospects and interest in the Russian oil and gas sector

The signing of Memoranda of Understanding (MOU) is a common practice to open the door for project-based activities under the Kyoto Protocol. An MOU is signed between the host country and possible investor countries. Multiple discussions are ongoing in Russia with potential investor/buyer countries that have expressed interest in Russian JI and GIS activities. Bilateral agreements regulating and facilitating JI activities were discussed between Russia and Canada, Japan and the European countries (Finland, Austria, Germany, Italy, Spain, France, Netherlands, Sweden, and Denmark).

Notwithstanding, Russia had still not signed any co-operation agreement with potential buyers or investors at the time of publication of this book. This being said, MEDT officials publicly reported that Russia has established common understanding with interested countries on the principles of such agreements. These agreements would be expected to determine the sectors of principal interest for both parties, necessary capacity building activities, and the project selection criteria, including principles for the additionality test. Several MOU agreements could be signed in 2006.

Foreign private-sector buyers are also expressing a growing interest in Russian JI projects. These are mostly large companies interested in projects in the energy sector. In the oil sector, gas flaring reduction projects are promoted by the World Bank Global Gas Flaring Reduction Public-Private Partnership (see Box 10).[80] Russian oil companies that have growing international profiles understand the importance of participating in GHG emission reduction efforts (over and above energy-saving and efficiency considerations). This is an important signal for international investors. They also take into account the possible marginal impact of carbon finance on investment decisions. However, successful implementation of gas flaring reduction projects is strongly dependent on structural limitations of the gas market, namely on more reliable third party access to Gazprom's transmission pipelines in Russia.

It is still unclear how Gazprom views JI. However, the company has already gathered operational experience during the pilot phase of AIJ through collaboration with Ruhrgas, which is considered a rare positive experience of the Russian pilot phase.[81] This project is viewed by several experts as an indication of Gazprom's preference to co-operate with its foreign business partners (suppliers and/or consumers) versus attracting new investors/buyers for its JI projects (Mielke et al., 2004).

On the one hand, the opacity of the Russian gas sector may contrast with the need to disclose information for determining baselines/additionality for JI projects. On the other hand, given high investment needs in Russian gas infrastructure (IEA, 2002, 2003, 2004) and the high capital intensity of most projects in the gas transmission sector, carbon revenues may have only a marginal impact on the economics of projects. Thus, priority may be given to JI projects in existing long-standing partnerships, rather than by explicitly carbon-driven projects (Platonova, 2005). We discuss this issue further in Chapter 3 on the basis of a typical compressor unit replacement project in the Russian gas transmission sub-sector.

Like RAO "Unified Energy Systems (UES) of Russia" (the largest power holding in Russia), Gazprom established a Carbon Agency in 2005 to screen possibilities for JI projects in the gas industry and to follow the development of national JI guidelines.[82] It is too early to say how effective the Carbon Agency will be. However, Gazprom has clearly the ability to match the leading position of RAO UES in the area of Kyoto-related activities in Russia.

---

80. GGFR (2003), personal communications with representatives of the GGFR.
81. See Chapter 3 for details on this project. The first phase of this project was implemented in 1997-99, with the potential for re-application in other parts of Gazprom's transmission system.
82. Discussions with Gazprom's officials in 2005.

The potential for projects to reduce GHG emissions in Russia's gas distribution sub-sector has also generated growing interest from international investors (Japan, Canada, Denmark and the World Bank). Moreover, gas distribution companies, such as Rosgazifikatsia (the state-owned company managing about 25% of the distribution facilities in Russia), are showing a clearer interest in the development of Kyoto-related activities, namely in the framework of JI. In fact, these distribution companies are viewing the Kyoto-related framework as a new opportunity to attract investment in energy efficiency given the current limited economic attractiveness of implementing such projects under business-as-usual conditions (see Chapter 4).

III. GHG EMISSION REDUCTIONS IN THE GAS TRANSMISSION SECTOR

Russia's high-pressure gas transportation system, the second largest in the world, transported 687 bcm in 2004 (Gazprom, 2005b). The linear part of the transmission system encompassed 153 300 kilometers of high-pressure trunk pipelines. The average distance that gas travels for domestic Russian consumers is about 2 400 km and 3 400 km for European consumers. There are about 260 compressor stations with more than 4 000 gas pumping units ensuring the necessary pressure to keep the gas flowing over these long distances. The system's size and the high volume of throughput make it one of the largest gas consumers in Russia.[83]

Worldwide, natural gas transmission systems are considered to produce large amounts of GHG emissions – encompassing energy-related emissions (from fuel consumption at compressor stations) as well as fugitive emissions (operational and involuntary releases of natural gas). Russia's Unified Gas Supply System (UGSS) is no exception. Gazprom which owns and controls the UGSS, attributes 98.5% of its CH_4 emissions to its gas transmission system.[84]

In this chapter we describe the current state of Russia's gas transmission system. We focus namely on its linear part or the transmission pipelines, and on the large stock of compressor stations. We identify the main sources of GHG emissions in the system and provide rough estimates of their level, focusing particularly on CH_4 emissions.

In general, estimates of energy-related CO_2 emissions are relatively straightforward to obtain using data on the energy efficiency of compressor stations as well as operational data. Estimates of fugitive emissions are less straightforward, due to the lack of specific emission factors by source. The best evaluation of CH_4 emissions is obtained by on-site measurement programmes, but these are costly to implement and their results have limited value in terms of extrapolation along the whole transmission system. This chapter presents an assessment of the level of GHG emissions along the Russian gas transportation system, based on the following sources:

■ Gazprom estimates, although extremely useful, are often difficult to compare with other gas transmission systems.

---

83. For the purpose of this study, underground storage facilities, although part of the transmission system to adjust gas flows to changing demand, are not included.
84. Gazprom's Environmental Report for 2004 includes only the set of emissions which are deemed atmospheric pollutants by Russian environmental law. As such the Report does not include all sources of GHG emissions.

■ Joint studies between Gazprom and foreign organisations and companies that provide emission factors for specific components of the Russian gas transmission system. However, their limited number and the relatively narrow focus of measurements make it difficult to extrapolate results across the whole system.

■ Comparisons with other countries with similar gas transmission systems, such as Canada and the United States, but unfortunately these are very limited in number and exact comparability.

Later in the chapter we describe the main options available to reduce GHG emissions along the linear part of the gas transmission system, and at compressor stations, based on Gazprom proposals and international experience and practice. Finally, we assess the potential role carbon finance can play in the refurbishment of compressor stations. Carbon finance could bring a double dividend based on the value of CO_2 and CH_4 emission reductions on top of the value stream generated by the saved gas from energy-efficiency improvements.

THE CURRENT STATE OF RUSSIA'S GAS TRANSMISSION SYSTEM

The linear part of Russia's gas transmission system

Russia's transmission pipeline system is the second largest in the world after the United States (Gazprom, 2004a). It is controlled by Gazprom, the Russian state monopoly, as a Unified Gas Supply System (UGSS) and consists of 9 corridors and 22 pipeline systems. Most of the system was built between 1975 and 1990, when the massive increase in gas production from West Siberia occurred. As shown in Table 9, pipelines built between the late 1960s to the present were of increasingly large diameter and able to withstand higher working pressure.[85] Most of the export pipelines are more recent and have technical parameters superior to those of domestic pipelines.

Table 9   Development of Russia's Unified Gas Supply System (UGSS)

Period	Diameter, mm	Pressure, MPa	Share of the system
Before 1960	325-530; 720-1 020	5.5	19%
	325	2-2.5	20%
1960-1968	1 020	5.5	11%
1968-1972	1 220	5.5	17%
1973-curr.	1 420	7.5	33%

Source: Renaissance Capital (2002) estimates based on Gazprom's Statistical Yearbooks.

---

85. Currently, pipelines with a diameter greater than 1 020 mm represent more than 60% of the total stock (*Energy Security of Russia*, 2005).

Large diameter pipelines (1 420 mm) with a pressure of 7.5 MPa make up the largest share of pipelines in Russia's UGSS.[86] They tend to be made primarily of steel covered with bitumen for insulation. Only a small fraction of Russian pipelines (2% in 1998) are made of polyethylene.

The technical operational lifespan of trunk pipelines is between 25 and 40 years. The fact that almost 60% of Gazprom's trunk pipeline system is over 20 years old (see Table 10) reflects the under-investment in this system over the 1990s up until 2002. This ageing process leads to increasing operational problems and the need for more intensive monitoring and repair. Ageing can also aggravate corrosion, especially in large diameter pipelines, thus increasing the risk of accidents (see Figure 15).

Table 10   Age structure of Gazprom's transmission pipelines

Years in use	Share of Gazprom's transmission pipeline system	
	2003	**2004**
0-10 years	14%	11%
11-20 years	38%	31%
21-35 years	30% (for 21-33 years)	41%
Over 35 years	n.a.	17%

Source: Gazprom, 2003, 2004a.

Figure 15   Accidents at transmission pipelines, 1991 to 2000

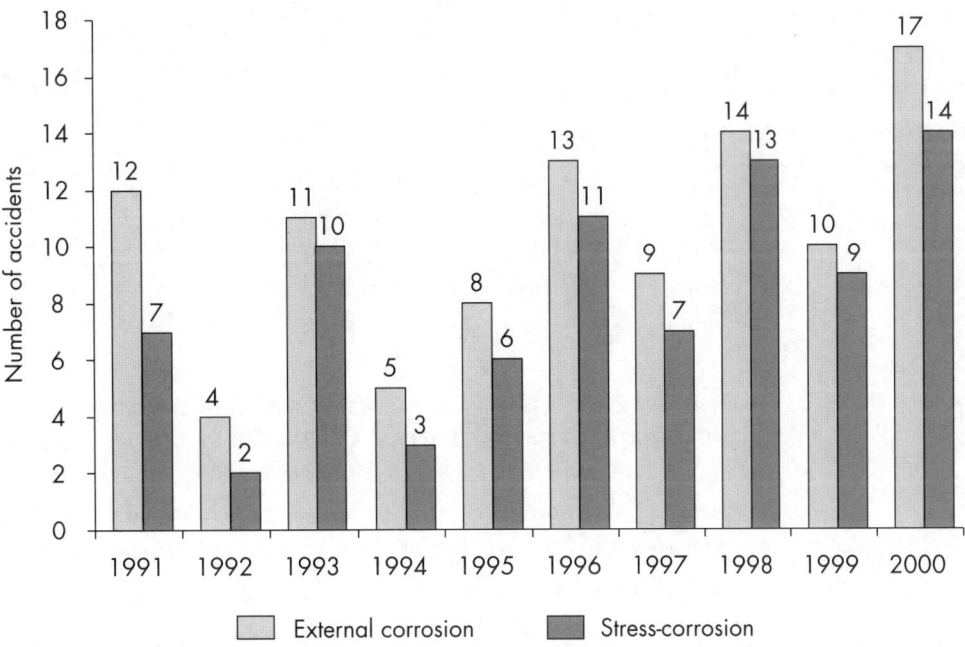

Source: Odishariya et al., 2003

---

86. In North America, average working pressure of pipelines is between 6.5 and 7.5 MPa (*Energy Security of Russia*, 2005).

The number of accidents per year is a key indicator of the reliability of the linear part of the transmission system. According to experts of VNIIGAZ, Gazprom's key scientific institute, the rate of diagnostic work from 1991 to 2001 did not keep pace with the rate of ageing of the system (Leontiev and Stureiko, 2003). During this period, the number of accidents due to stress-corrosion, a major cause of accidents, significantly increased (see Figure 15).

The ageing of pipelines also leads to reduced pipeline pressure ratings or reduced volumes of natural gas throughput. In 2003, 16% of Gazprom's transmission pipelines (24 600 km) had reduced pressure ratings and thus resulted in limiting the volume of gas which the system could transport. "Bottle necks" in the system limited throughput in and beyond the actual "weak" sections. This also causes higher energy consumption within the transmission system as gas has to be compressed to a higher pressure for onward transmission beyond the weak sections in the pipeline system. Therefore the amount of natural gas consumed increases in the transmission process to final consumers, domestic or foreign.

In 2002, experts of VNIIGAZ estimated that the operational capacity of Gazprom's transmission system was 60 bcm (almost 10%) less than the system was designed to carry, resulting in transportation of about 60 bcm less than its rated capacity (Leontiev and Stureiko, 2003; Pravosudov, 2004a). This reflected an increasingly serious trend, given that the level of reduced throughput in 1991 was estimated at only 24 bcm. This shows the importance of annual pipeline maintenance and refurbishment programmes in ensuring sustainable and reliable gas supplies for domestic and export markets.

Gazprom's reconstruction programmes were largely underfinanced in 1991-2001 (see Figure 16). As a result, throughput capacity became more and more limited. Investment needs trebled between 1991 and 1996 and more than doubled again by 2002. Gazprom is nearing the end of its third five-year Comprehensive Programme for Reconstruction and Repair of its transmission system. According to various sources, including the Head of Gazprom's Transportation Department, 237 billion roubles (about USD 8 billion) were to be invested in this programme during the five-year period to 2006 (Budzulyak, 2004). By 2004, Gazprom reported it was surpassing its annual targets, having already invested 80% of the projected funds.

Unlike the two earlier 5-year programmes, Gazprom appears to be meeting the investment targets set out in its 2002-2006 Reconstruction Programme (Budzulyak, 2004; Kirillov, 2005). It increased funds for reconstruction and capital repair to about USD 2 billion per year (see Figure 17). The goal of the programme is to maintain the reliability and security of the system as well as the viable functioning of its transmission system. The programme has sought to:

■ Increase Gazprom's operational capacity by 35 bcm by 2006 (out of the total estimated limited capacity of 60 bcm per year).

■ Reduce gas consumed at compressor stations by 5 bcm by 2006 (or 12% of gas consumption by Gazprom transmission pipelines in 2004).

Figure 16   Funds allocated and actually invested over
Gazprom's five-year Reconstruction Programmes from 1991 to 2006

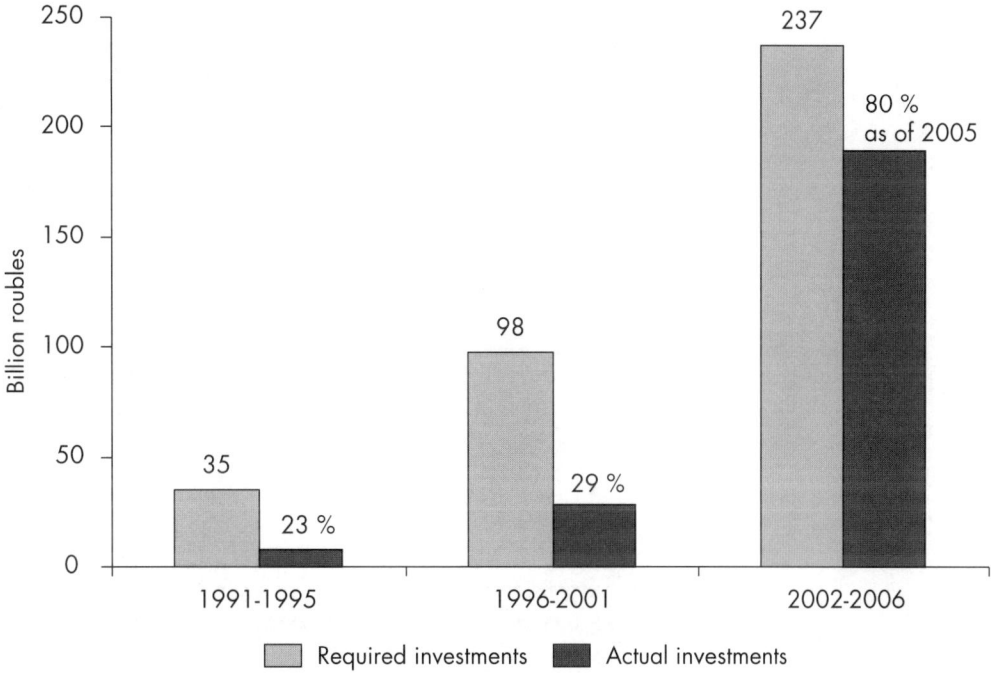

Sources: Leontiev and Stureiko, 2003; IEA estimates for investments in 2002-04.

Figure 17   Gazprom investment in the reconstruction and overhaul of the transmission
system, 2000-04

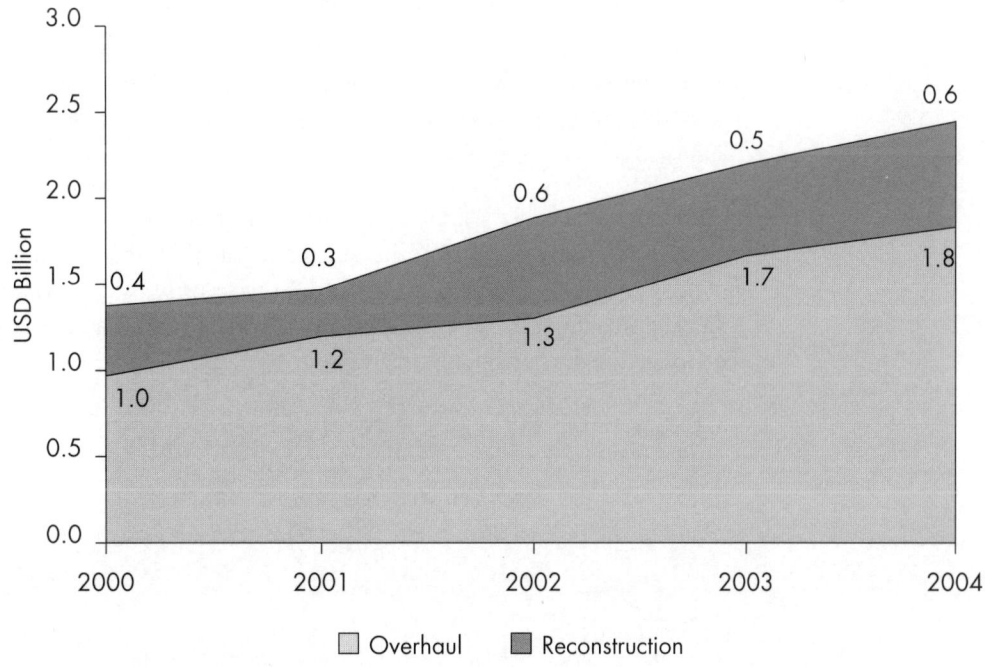

Source: Kirillov, 2005.

Figure 18   Implementation of Gazprom's pipeline flaw detection programme, 1993-2004

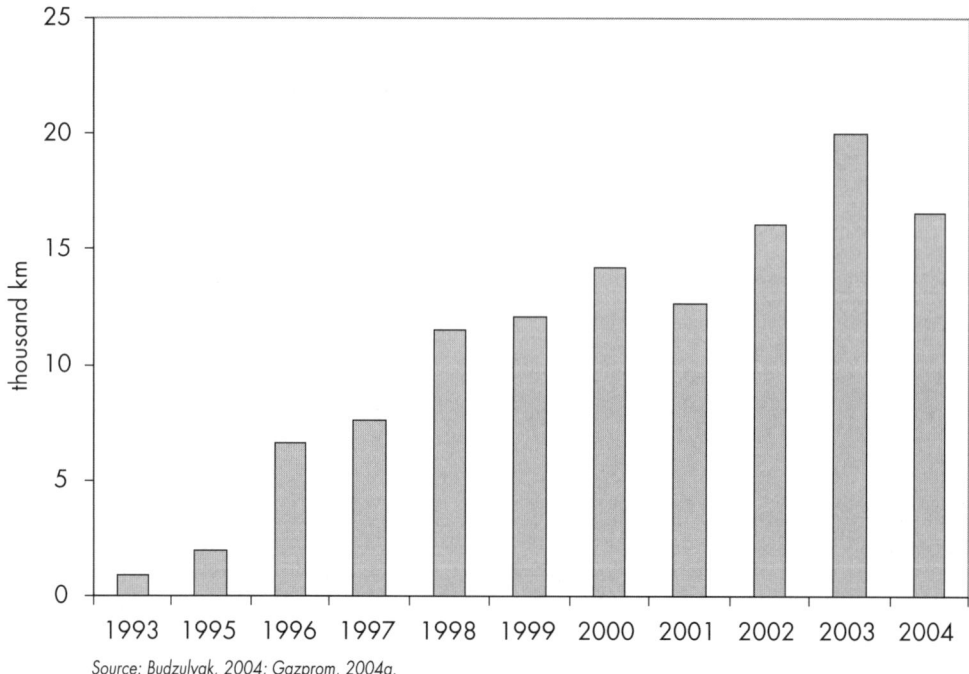

Source: Budzulyak, 2004; Gazprom, 2004a.

Unfortunately, there is no detailed information on Gazprom investments in its pipeline refurbishment programme after 2004. Information available on its Web site lists only the various activities undertaken in 2005 with no reference to the progress it is making in achieving its programme goals in terms of increasing operational capacity.

Since 2002, the number of internal pipeline diagnostic measures and flaw detection programmes has also increased (see Figure 18). The annual rate of overhaul and reconstruction of pipelines for the entire network increased from 0.41% of the total length from 1990-2000 to roughly 1% in 2003-04.[87] As a result, the number of reported accidents decreased from 0.21 per 1 000 km in 2002 to 0.18 in 2003 (Odishariya *et al.*, 2003; Gazprom, 2003).

Despite this decline, "Gaznadzor"[88] experts argue that without an increase in reconstruction and overhaul of the gas transmission system, the number of accidents could dramatically increase. This could lead to security risks for Gazprom as well as the whole country, as more than 44 000 km of transmission pipelines and facilities are located in densely populated areas.

In the future, this will become increasingly important as Russia's transmission system will have to transport bigger volumes of natural gas to meet both its domestic needs and increasing exports. Gazprom has used its limited transmission capacity as a key reason in the past to deny third party access to its system. Although it argues that

---

87. The overhaul of the pipeline network averaged about 620 km per year from 1990 to 2000, increased to 1 350 km in 2003 and 1 913 km in 2004 (*Gazprom Annual Reports*, 2003, 2004).
88. Gaznadzor is a subsidiary of Gazprom responsible for the verification and monitoring of the rules of operation and construction of gas pipelines and facilities of the UGSS, and for the audit of energy use.

access by independent gas producers has increased over time, much of this has been given to Gazprom-affiliated companies shipping Central Asian supplies to Ukraine. Meanwhile, investment by independent producers of natural gas, Gazprom's potential competitors in Russia's upstream gas sector, has been hampered by unreliable and unpredictable access to Gazprom's transmission system.

This raises concerns for longer-term energy security for Russian and European consumers if currently-planned new export pipelines are filled with more expensive Russian natural gas due to the lack of competition within Russia, or if under-investment in the domestic transportation system erodes the ability of Gazprom to meet increasing demand. Kyoto flexibility mechanisms could play a marginal role in helping to provide a stimulus to accelerate the rate of pipeline refurbishment and to increase throughput capacity. The mechanisms could help enhance investments in Russia's transmission system and thereby help to support a burgeoning competition in its upstream gas sector.

Compressor stations

The UGSS includes 263 compressor stations (see Box 5) driven by 4 067 gas pumping (compressor) units, which are powered by engines with a total capacity of 44.2 GW (Gazprom, 2004). In 2001, more than three quarters of Gazprom's compressor units were based on gas turbine drives and these supplied 85% of the UGSS power needs.[89] Many of these units are obsolete or past their technical lifespan, including 13% operating for more than 100 000 hours and another 49% more than 50 000 hours (see Table 11).[90] The structure of Russia's stock of compressor units by type and age is given in Annex 3.

The capacity of gas turbine engines ranges from 2.5 to 25 MW with unit efficiency factors ranging from 23 to 35%. From 1996, replacement and reconstruction of compressor units increased the average efficiency of the stock of compressor units from 27.6 to 28.2% in 2000 (Gazprom, 2001a). During this period, a total of 287 units were reconstructed with new drives with efficiencies in the order of 31-37%. However, this average efficiency remains significantly lower than that of western counterparts and considerably lower than that of the best available technology (BAT) (see Annex 4). For example, the stock of Canadian compressor units has an average efficiency of 35% (TransCanada, 2002).

Table 11

Hours of operation of Gazprom's compressor units

Hours of operation	Share of Gazprom's total compressor units
Less than 50 000 hours	38%
From 50 000 to 100 000 hours	49%
More than 100 000 hours	13%

Source: IBRC, 2003.

---

89. Electric drives make up 14%, while piston drives supplied less than 1% and are used primarily for gas injection into underground gas storage (Gazprom, 2001a). In this book mainly gas-driven compressor units are considered as they represent the bulk of energy consumption by transmission pipelines.
90. Over half of the power capacity of gas turbines in Russia is from the old and low efficient units of the 1st and 2nd generation. These turbines represented about 43% of the capacity of Tyumentransgaz in 2001 (IBRC, 2003) through which about 80% of Russia's gas passes (Maslennikov, 2004).

Box 5   Compressor stations along gas transmission pipelines

Natural gas moving through a pipeline is compressed to about 100 times the atmospheric pressure, which drives it along at speeds of up to 40 km/hour. The friction of the natural gas moving through the pipe results in a loss of pressure. Compressor stations are typically located every 75-150 km in Canada and the United States. In Russia, Gazprom aims to space compressor stations every 100-120 km. Compressor stations may have several compressor units, which are powered by engines (or prime movers), some of which are similar to jet engine turbines. These can be powered by electric motors, diesel or natural gas engines or turbines. Compressors are also used in the liquefaction of natural gas, at storage facilities, and to help inject the natural gas into high-pressure storage fields. The operational life of compressors tends to be typically around 50-60 years, while the lifetime of turbines (engines) is only 15-17 years, or 130 000 to 150 0000 hours.

Figure 19   Illustration of the functioning of a natural gas compressor unit

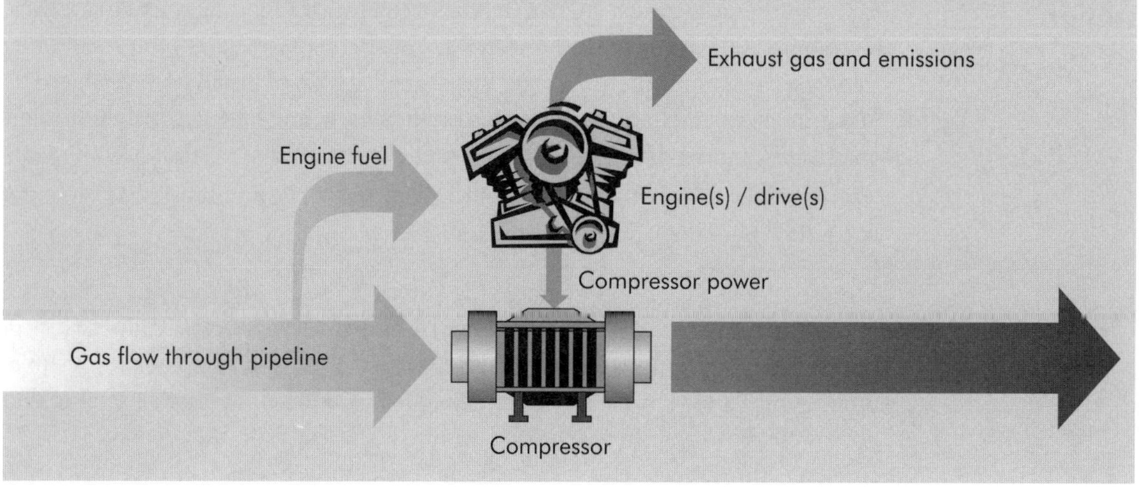

Exhaust gas and emissions

Engine fuel

Engine(s) / drive(s)

Compressor power

Gas flow through pipeline

Compressor

SOURCES OF GHG EMISSIONS IN THE GAS TRANSMISSION SYSTEM

Energy-related CO$_2$ emissions from compressor stations

According to Gazprom, annual consumption of gas by its compressor stations ranges between 42-45 bcm (Kirillov, 2005). IEA calculations based on overall Gazprom gas consumption figures and the structure of its gas consumption over 1997-99, estimate the level of gas consumed by its compressor stations at about 40.8 bcm or 80-85% of Gazprom's total gas consumption (see Table 12).[91] Although this simple extrapolation does not take into account the impact of possible improvements in the efficiency of compressor stations since 1999, it is much in line with the volumes reported by Gazprom (*Energy Security of Russia*, 2005).

---

91. Losses include technological losses and errors of meters (Gazprom, 2001b).

Table 12   Gas consumption by Gazprom's gas transmission system, 1997-2004

	Gazprom data				Estimates				
	1997	1998	1999		2000	2001	2002	2003	2004
			bcm	%					
Gazprom's gas consumption, bcm including:	47.6	50.9	50.7	100%	47.1	45.7	47.7	51.3	52
Combustion by compressors	36.5	39.8	39.8	79%	37.0	35.9	37.4	40.3	40.8
Technological needs	4.6	4.4	4.3	9%	4.0	3.9	4.1	4.4	4.4
Losses	6.5	6.7	6.7	13%	6.2	6.0	6.3	6.8	6.9

Sources: 1997-99 figures from Gazprom (2001b); IEA calculations based on Gazprom (2004a).

Energy consumption by the gas transmission system

The energy intensity of Russia's gas transmission system is 30-60% higher than comparable foreign systems. [92] This is mainly due to:

■ Russian natural gas having to travel long distances.

■ The relatively low energy efficiency of compressor units leading to 10-20% over-consumption of gas according to Gazprom.

■ The insufficient power capacity of booster compressor stations leading to increasing energy consumption on the main compressor stations. A deficit of each kW of power capacity on the booster compressor stations results in an additional use of 4kW on the compressor stations along the transmission system.

■ The initial design of the system, influenced by low energy prices during Soviet times, mainly aimed at saving pipe metal and limiting the number of compressor units.

■ The compressor ratio of gas in Russia (1.45) is higher than in comparable foreign systems (1.3 to 1.35) and thus results in higher energy needs.[93] The lower compression of gas in foreign systems is due to the larger diameter of pipelines and lower internal pipe roughness.[94] According to Gazprom, the energy intensity of the Russian gas transmission system actually decreased in 1990-2000, mainly due to lower production. Since 2001, Russian gas production, as well as transit or imports from Central Asian countries, has been increasing. Thus, unless efficiency enhancements are made, the transmission system's energy intensity will continue to increase once again.

Energy-related CO_2 emissions from compressor stations

Until recently, there has been less focus on CO_2 emissions associated with the combustion of natural gas used to drive gas turbines at Russian compressor stations compared to those of its foreign counterparts. Estimates can be made based on operational data on fuel gas usage, service schedules, machine start-ups and running times, etc., as well as from typical design parameters for Russian machinery (Wuppertal Institute, 2005).

---

92. *Energy Security of Russia*, 2005; Gazprom, 2001b.
93. The compressor ratio is the difference between the inlet and outlet gas pressure at compressor stations.
94. Gazprom estimates 15-20% over-consumption of energy due to rougher pipeline surfaces (Gazprom, 2001b).

Alternatively, the most straightforward estimates can be determined applying emission factors for natural gas combustion to the volume of fuel gas used by the compressor.[95] As we show in Annex 1, based on IPCC guidelines, the combustion of 1 m³ of natural gas results in roughly 2kg of CO_2. According to estimates from Table 12, in 2004, energy-related CO_2 emissions along Russia's gas transmission system were in the order of 82 $MtCO_2$. This represents about 47% of GHG emissions in the transmission system (see Table 14).[96]

Methane emissions in Russia's gas transmission system

In principle, every component of the gas transmission system can lead to fugitive methane emissions. Two types of fugitive emissions are generally distinguished: i) unplanned or involuntary emissions from "leaks" or possible technical problems or accidents ("blowouts"), and ii) operational discharges ("venting").

Unplanned or involuntary emissions from "leaks" or "blowouts"

Fugitive emissions

Leaks can occur at, in, and around various fittings (valves, bolted assemblies, joints and flanges), through holes in pipelines as well as from safety devices (vents), for example:

- Fittings may leak because of their design and direct association with the gas-bearing system of the compressor station and transmission line. The number of leaks in transmission pipelines also depends on the type of material used – protected steel or plastic.

- Seal valves upstream of the vents may also leak, allowing gas to escape from the vents.

- Gas leaks can also be due to breakdowns, i.e. pipe fractures or accidents. Blowouts are generally due to a dangerous build-up of system pressure that results in gas being released into the atmosphere.

- Methane may also be released from isolation valves and/or blow-down valves from off-line compressors.

Operational/technological discharges or "venting"

Vents can discharge gas into the atmosphere under controlled conditions, *e.g.* for venting purposes. Venting occurs in all parts of the production and supply chain.[97] The standard safety practice is to "blow down" or "vent" the high-pressure gas into the atmosphere during equipment decompression. This is carried out for safety reasons when compressors are taken off-line for maintenance, repairs and emergencies. A common decompression technique is to block off the affected segment and then vent the CH_4 that is present into the atmosphere. This is referred to as a blowdown.[98]

---

95. See Annex 1 for details on calculation of emission factors.
96. Wuppertal Institute (2005) concluded that CO_2 emissions from gas combustion at compressor stations, represented about 60% of GHG emissions of the two Gazprom export pipelines which were inspected.
97. Up to several tonnes of methane can be released at one moment.
98. Natural Gas STAR (2004a) indicates that on average in North America, a single blowdown results in the release of about 0.42 Mm³ or 6.4 $ktCO_2e$ of gas to the atmosphere.

However, there is no clear distinction between leaks and vents, given the difficulty in classifying fugitive emissions as intentional or unintentional. In what follows, we use the common term "fugitive emissions".

Estimations of CH$_4$ emissions of gas transmission systems

Guidelines for methane emission inventories

It is extremely difficult to establish accurate emission factors for individual companies given the large number of potentially leaking components of transmission systems. In most cases methane emission inventories are based partly on a company's own measurements as well as on emission factors derived in past surveys and studies. Companies can evaluate their fugitive emissions using various methods: balances, calculations using standards and codes, or the instrumental method with periodic on-site measurements (Remizov *et al.*, 2000). The instrumental method using CH$_4$ detectors is more reliable but expensive and time-consuming.

Several programmes have been developed to assist oil and gas companies to establish an inventory of their methane emissions in order to be able to calculate environmental benefits from their emission reduction activities:

■ The *Petroleum Industry Guidelines for Reporting Greenhouse Gas Emissions* (2003) jointly developed by the International Petroleum Industry Environmental Conservation Association (IPIECA), the Oil and Gas Producers (OGP), and the American Petroleum Institute (API).

■ The GHG protocol including *A Corporate Accounting and Reporting Standard* (revised edition, 2004) and *The GHG Protocol for Project Accounting* (November, 2005) developed by the World Resources Institute (WRI) and the World Business Council for Sustainable Development (WBCSD).

■ The *Preferred and Alternative Methods for Estimating Air Emissions from Equipment Leaks* (November, 1996) and *Preferred and Alternative Methods for Estimating Air Emissions from Oil and Gas Field Production and Processing Operations* (September, 1999) developed and accepted by the United States Environmental Protection Agency (US EPA) (Natural Gas STAR, 2004b).

■ The guidelines on *Calculating Greenhouse Gas Emissions* developed by the Canadian Association of Petroleum Producers (CAPP, 2003).

■ The study on *Methane Emissions from United States Natural Gas Systems* by US EPA and Gas Research Institute (1996). The emission factors are presented by segment, emission type and source (equipment and components of gas systems).

The IPCC Guidelines for the preparation of national emission inventories provides several specific regional CH$_4$ emission factors for gas transmission systems (excluding the FSU).[99] However, the use of fugitive emission factors for various types of

---

99. IPCC (2000) contains CH$_4$ emission factors based on North American data. Other emission factors can also be found in the approved CDM methodology for gas flaring reduction projects (CDM EB, 2005c).

equipment and technologies have two major disadvantages: i) it is a data-intensive process requiring large up-front investments in data collection; and ii) there are considerable difficulties in uniformly using these emission factors, given that the maintenance and operational structures could differ significantly from the "typical" value. For this reason, despite the numerous studies assessing various methane emission factors, many Russian experts consider that the significant difference in age and design of Russian facilities makes the use of foreign emission factors difficult in Russia (Kokorin, 2005).

Thus, continued measurement and data collection of different components of the gas transmission system is essential for developing appropriate activity data and complete databases of emission factors for Russia.

A comprehensive and verifiable methodology of CH_4 emission inventory is also necessary for calculating emission reductions and assessing their potential value on the carbon market. Such a methodology is also needed to establish baseline scenarios and to estimate a project's emission reductions. IEA member country gas companies and Gazprom have a growing interest in developing an international methodology of CH_4 emission inventory, taking into account the specificity of different gas systems.[100]

CH_4 emission measurement programmes along gas transmission systems

Several CH_4 emission measurement programmes along Russia's gas transmission system were implemented over the last ten years by Gazprom (including VNIIGAZ) in co-operation with various other gas companies, such as Ruhrgas, Gaz de France and Sumitomo of Japan, as well as with the United States EPA and various independent experts (Remizov et al., 2000). These programmes focussed on different components of the transmission system. In order to obtain a full picture of the main sources of emissions, the following studies are considered:

■ The Gazprom & Ruhrgas study based on a common measurement programme of compressors, pipelines and gas production and processing facilities conducted in 1996-97 (reported in Dedikov et al., 1999). The details on measurements on gas production and processing facilities are presented in Annex 5.

■ The Gazprom & US EPA (1996) study which included measurements at compressor stations.

■ The Gazprom & TransCanada CH_4 emission reduction project implemented in 2001 at two compressor stations (1 GW) including measurements of valve leaks (Venugopal, 2003).

■ The Wuppertal Institute (2005) measurement campaign was implemented in cooperation with Max-Plank-Institute and on behalf of E.ON Ruhrgas. The selection of sites was made jointly with Gazprom and its scientific institute VNIIGAZ.

---

100. At the European level, the development of such methodology is a focus of the Eurogas-Marcogas working group including Gaz de France, SNAM Rete Gas, OMV Gas, Transco, Ruhrgas, Italgas and Figas.

Linear part of the transmission system

The Gazprom & Ruhrgas study was undertaken in 1996-97 focussing on the transmission system pipelines of the Tyumentransgaz and Volgotransgaz (Uzhgorod Corridor) systems. Based on an inspection of 630 km of the Volgotransgaz pipelines, the rate of emissions amounted to 8 200 m³/km/year. Emissions due to venting for the purpose of repairs accounted for almost 60%, leaks represented about 33% of total emissions, and the remaining 9% was due to ruptures (see Annex 5 for details).

In 2003, the Wuppertal Institute (2005) undertook a separate study to measure the emissions along just over a third of the length of the export trunk pipeline system of the Central and Northern Corridors, covering a distance of 3 376 km and 3 075 km, respectively, linking the production regions of West Siberia to Germany and Western Europe.[101] The CH_4 emissions along the linear part of these pipeline systems were estimated at 6 458 m³/km per year – 20% lower than the earlier estimates made by the 1996-97 Gazprom & Ruhrgas study. Despite this difference, the structure of emissions of the 2003 study reflected, to a large extent that of the earlier study. Gas vented before maintenance and repair accounted for over 58% of emissions, leaks represented another 38%, while ruptures contributed only 4%.[102] The Wuppertal Institute study reflects improvements in diagnosis and preventive repairs by Gazprom over the interim period given the significant reduction in emissions due to accidents (half the level estimated by the 1996 Gazprom & Ruhrgas study).

Compressor stations

Compressor stations can have up to 2 500 different components, most of which are liable to intentional or unintentional leaks. Studies by the US EPA & GRI (1996a), however, show that the largest leaks at compressors are due to relatively few components. From 1995 to 2005 various measurement programmes were undertaken at compressor stations in Russia. Results are summarised in Table 13 and presented in more detail in Annex 6. Unfortunately, due to different measurement techniques and different sets of measured components, estimates of emissions from different sources may not be fully comparable across various measurement programmes or countries.

The Gazprom & US EPA study included measurements of CH_4 emissions at four compressor stations, with a total capacity of 924 MW located near Saratov and Michurinsk (see Annex 6). The measurements focussed on a limited number of components and further data collection was necessary to obtain more precise emission factors. Extrapolation of the results over Gazprom's entire stock of compressor units provides an estimation of CH_4 emissions of 2 bcm per year. Given the narrow focus of the study, its results should be considered as lower-bound estimates.

This study, although limited, was a useful first step towards identifying emission sources for priority mitigation. It showed that leaks in Russia have a high degree of

---

101. These pipelines consist of 4-6 parallel pipes. The total pipeline length is 22 000 km for the Central Corridor and 12 000 km for the Northern Corridor.
102. These emissions were calculated according to the accident statistics provided by Gazprom.

Table 13   Estimates of methane emissions by Russian compressor stations

Reference	Methane emissions	Measurements
Gazprom & Ruhrgas (Dedikov *et al.*, 1999)	Kazym station: 75 000 m³/MW/y	Operational vents and unintentional leaks measured in 1996-97 in Kazym station (1971-77) and Upper Kazym station (1983-97)
	Upper Kazym station: 53 000 m³/MW/y	
Gazprom & TransCanada (Venugopal *et al.*, 2003)	21 364 m³/MW/y	Measurement of valve leaks in 2001 in Pochinki and Torobeevo stations
Wuppertal Institute (2005)	49 418 m³/MW/y	Operation-related emissions (5 227 m³/MW/y) and unintentional leaks (44 191 m³/MW/y)

concentration by source, consistent with leak patterns in the United States, and that 85% of emissions came from vents mainly during compressor downtimes and before maintenance and repairs.

In 1996, the Gazprom & Ruhrgas study measured emissions at two compressor stations: Kazym ("old") and Upper Kazym ("new").[103] Not surprisingly, the older station showed higher CH_4 emissions, estimated at 75 000 m³/MW per year, compared to the newer station's emissions of 53 000 m³/MW per year. This difference was attributed to leaks from compressor seal oil systems. Vents were the main cause of CH_4 emissions representing more than 50% for both stations. Vents included emissions during repairs, start-up and depressurisation of compressor units, degassing of seal oil, and the incomplete combustion of methane in gas turbines.

The Gazprom & TransCanada project to reduce CH_4 emissions was implemented in 2001 at two compressors stations, Pochinki and Torbeevo (including 71 compressor units with total capacity of 1 GW). The project focussed on valve leaks and detected 1 500 leaking sources (Venugopal *et al.*, 2003). The average rate of leaks per MW was estimated at 21 364 m³/MW/y. Venugopal *et al.* (2003) extrapolated this emission factor over Gazprom's entire compressor stock (40 GW in 2001) and obtained annual CH_4 emissions from Gazprom compressors of 0.9 bcm/y or 14 $MtCO_2e$/y. These results should also be considered as lower-bound estimates as the project focus was only on valve leaks.

The Gazprom & TransCanada project resulted in a reduction of about 10 Mm³ of gas or 0.15 $MtCO_2e$ of CH_4 emissions per year (TransCanada, 2002). Based on this, Venugopal *et al.* (2003) estimate that Gazprom could reduce valve leaks by half (in the order of 0.4 bcm/y of CH_4) across its entire system and reduce emissions by about 7 $MtCO_2e$ per year.

The Wuppertal Institute study along the Central and Northern Corridors included compressor stations with 50 compressor units (534 MW) and gate valves (see Annex 6

---

103. Both compressor stations use typical equipment (*i.e.* GT-6-750, GTK-10, GPA-Z-16), thus are considered representative in the Gazprom & Ruhrgas study.

and Table 13). Compressors of the Central Corridor are newer and have a higher capacity than those of the Northern Corridor. The study concluded that unintentional leaks from fittings and vents were the main source of CH_4 emissions accounting for 80% of total. The resulting emission factor of 49 418 m³/MW per year was consistent with those estimated at the Upper Kazym ("new") compressor stations in 1997 in the Gazprom & Ruhrgas study.

Extrapolation of CH_4 emission factors over Gazprom's entire transmission system

The large scale of gas networks, especially in Russia, makes it impossible to investigate all stations and sections of transmission pipelines with the instrumental method (Wuppertal Institute, 2005). Generally, companies make measurements on a selection of gas facilities, which reflect the conditions of a specific part of, or in the best case, the whole gas transmission network. Results can be used to make extrapolations, with obvious shortcomings and uncertainties the larger and heterogeneous the system.

For example, the 2003 measurement campaign of the Wuppertal Institute described above, estimated CH_4 emissions of two export corridors of the Russian gas transmission system. The goal was to include compressor stations of both export corridors, operated by different Gazprom subsidiaries, and exposed to different geographic locations and infrastructures.

The Wuppertal Institute extrapolates its results over Gazprom's gas export transmission system and estimates fugitive emissions at approximately 0.7% of the volume of gas exports (the range from 0.4 to 1.6 % with 95% certainty). This is comparable to the estimated amount of gas lost from the United States' gas transmission sector estimated in 1992 at 0.5±0.19% of gross gas production (US EPA and GRI, 1996a).[104]

Based on the Wuppertal Institute study, methane emissions from the export part of the Russian gas transmission system are evaluated at 1 bcm in 2004 (the range from 0.6 to 2.3 bcm). However, as only a relatively small statistical sample was studied, it is not sufficiently representative of the whole system. Futhermore sampled facilities are newer and in better condition than the system average. For this reason, the Wuppertal Institute (as well as Ruhrgas experts) considers the results of extrapolations across the whole transmission system unreliable.

Despite the shortcomings of extrapolations, Gazprom experts use the results from the Gazprom & Ruhrgas study to estimate the volume of total gas lost in the atmosphere across Gazprom's entire system (Dedikov *et al.*, 1999). This extrapolation estimated that CH_4 leaks from Gazprom's transmission system were in the order of 1±0.5% of the natural gas produced. Results were broken down by segment:[105]

■ Production and processing emissions represented 0.1±0.05% of total produced gas.

---

104. In the United States, CH_4 emissions in the natural gas industry (including the distribution sector) were estimated at 1.4±0.5% of gross natural gas production in 1992 (US EPA and GRI, 1996).
105. The uncertainty of emission factors obtained by the Gazprom & Ruhrgas study is ±50% (Popov, 2001).

■ Linear part of the transmission system, *i.e.* pipeline emissions – 0.2±0.1% of total throughput.

■ Compressor stations – 0.7±0.35% of total throughput.

According to Dedikov *et al.* (1999), Gazprom's CH_4 emissions along its gas production and transmission systems were of the same order of magnitude as gas systems in western industrial nations where methane emissions amount to 0.1-1.5% of gas produced or consumed (International Gas Union, 1997). Russia clearly stands out with emissions in the upper-end of this range. While these results were confirmed by the more recent Wuppertal Institute study, it focussed only on sections of export pipelines, the best maintained part of the system. The IEA considers that Dedikov *et al.* (1999) results are conservative estimates of CH_4 emissions in the Russian transmission sector. Considering the under-investment in pipeline refurbishment programmes over the 1990s to 2002, these results should still be considered as lower-bound.

Based on the studies, in 2004 CH_4 emissions along Gazprom's gas transmission system can be estimated at 6.2 bcm of natural gas or 92.8 $MtCO_2e$ (see Table 14). This represents about 1% of total gas throughput. However, more independent studies would be necessary to provide more comprehensive emission estimates.

The IEA estimates total GHG emissions in 2004 from Russia's gas transmission system at about 174 $MtCO_2e$ (see Table 14). This is based on the total volume of transported natural gas, including the gas production of independent gas producers. CH_4 leaks represent about 53% of the total and are subject to high uncertainty. Compressor stations account for about 90% of total GHG emissions of the transmission system and are the main source of fugitive CH_4 emissions and CO_2 emissions from combustion. There is great potential for compressor stations to reduce these emissions. However, even if CO_2 emissions can be reduced through more energy-efficient technologies, energy needs cannot be reduced to zero.

Table 14 Estimates of GHG emissions along Russia's gas transmission system in 2004

Estimates for 2004	bcm	$MtCO_2e$	GHG Emission structure
Volume of natural gas in transmission pipelines	687.4	–	–
Gas consumed by compressor stations	40.8	–	–
CO_2 emissions from compressor stations	–	81.6	47%
CH_4 emissions at compressor stations	4.8	72.2	41%
CH_4 emission rate, % of gas throughput	0.7%	–	–
CH_4 emissions along transmission pipelines	1.4	20.6	12%
CH_4 emission rate, % of gas throughput	0.2%	–	–
Total CH_4 emissions	6.2	92.8	53%
Total GHG emissions	**–**	**174.4**	**100%**

Sources: IEA estimates based on Gazprom data and results of the above-listed CH_4 emission studies.

Table 15   Classification of fugitive CH_4 emissions along transmission pipelines and at compressor stations

Methane emission factors	Low	Medium	High
Linear part of transmission system, tCO₂e/km/year	3	30	300
Gazprom & Ruhrgas study (1996-97)			121
Wuppertal Institute (2005)			95
Compressor stations, tCO₂e/MW/year	90	300	1500
Gazprom & Ruhrgas study (1996-97)			782 - 1107
Gazprom & TransCanada study (2001)		315	
Wuppertal Institute (2005)			729

Sources: Hanle (2003) derived from Altfeld et al. (2000); IPCC, 2000; IEA calculations.

To facilitate international comparisons, the *IPCC Good Practice Guidanes* (2000) provide default emission factors (see Table 15) for the gas transmission sector, which were proposed in 1995 by the International Gas Union (IGU). These emission factors reflect possible low, medium and high CH_4 emissions for facilities commissioned in 1995 or earlier (Hanle, 2003).

In Table 15 we compare these emission factors with the results of measurements in Russia. It is worth noting that this comparison is indicative of the level of emissions only, given the difference in design and parameters of the systems, maintenance practices, climatic conditions, and measurement methods. In addition, these factors do not take into account the improvements, which may have been realised by the gas transmission companies in order to reduce gas losses and fugitive emissions from 1995 to the present day. For example, according to TransCanada (2002), the CH_4 emission factors for Canadian pipelines were reduced by 42% from 1999 to 2001.

According to Table 15, fugitive emissions along the linear part of Russia's gas transmission system can be considered as medium bordering on high in comparison with the default emission factors of IPCC. Emissions at compressor stations border more on the higher level, even for facilities commissioned in the mid-1980s indicating a potential for further CH_4 emission reductions. At the same time, the significant differences in the estimations as indicated in the studies suggest that more effort is needed to determine proper emission factors for GHG inventories. Although of limited use for specific projects, these emission factors can provide information on the state of equipment in Russia and facilitate the development of a baseline for Kyoto-related project-based activities.

MAIN OPTIONS TO REDUCE GHG EMISSIONS

According to Gazprom (2001b), the gas transmission system can provide over 70% of potential energy savings of the whole gas industry through compressor replacement and the implementation of efficient maintenance and repair practices. These measures are attractive given the benefits from improvements in the operational capacity of the gas transmission system and the reduction in gas losses and own consumption. In other words, these measures would reduce costs, while at the same time increase the volume of natural gas available for sale on the domestic and/or export markets (and substitute more costly production). With the added bonus of benefits under Kyoto-related flexibility mechanisms, these measures are all the more attractive.

GHG emission reduction options considered by Gazprom to 2012

Gazprom prepared a set of GHG emission mitigation options to be implemented in the gas transmission sector in two phases: 2001-08 and 2008-12 (see Table 16). These options contribute directly to the implementation of Gazprom's Development Programme (2002-10), where Gazprom stresses the growing importance of measures to maintain and improve its operational capacity, viability, security and economic efficiency of the gas transmission system. The total estimated potential of GHG emission mitigation is 51.4 $MtCO_2$ per year by 2012 or roughly 30% of total GHG emissions of the gas transmission sector.

A set of GHG emission reductions options

In terms of energy (gas) savings, the replacement and modernisation of compressor units is the most attractive measure, providing 60% of the total expected gas savings (or about 15% of the 40.8 bcm of gas consumed by gas turbines in 2004). The use of modern technologies for maintenance and repair[106] will provide maximum GHG emission reductions (37%) and replacing compressors and implementing low compressor ratio gas transportation regimes will contribute to another 40% of total GHG emission reductions expected by Gazprom.

Table 16   Gazprom's expected reductions in GHG emissions in the transmission system, 2001 vs 2012

	Reduction of CH_4 emissions		Reduction of gas use (combustion)		Total GHG emission reductions	
	bcm	$MtCO_2e$	bcm	$MtCO_2e$	bcm	$MtCO_2e$
Replacement and modernisation of compressor units	–	–	6	11	6	11
Optimisation of the operational regimes using specific software	–	–	1.0-1.4	1.8-2.6	1.0-1.4	1.8-2.6
Low compressor ratio gas transportation regime	0.8-1.0	12-15	–	–	0.8-1.0	12-15
Use of modern maintenance and repair technologies	1.2-1.4	18-21	–	–	1.2-1.4	18-21
Cleaning of pipelines	0.4	6	–	–	0.4	6
Total	**2.4-2.8**	**36-42**	**7.0-7.4**	**12.8-13.6**	**9.4-10.2**	**49.2-55.2**
Mean value of Gazprom estimates	2.6	38.2	7.2	13.2	9.8	51.4

Source: Energy Security of Russia, 2005.

---

106. Further gas savings are also expected through the replacement and modernisation of compressor units in the form of CH_4 emission reductions due to the elimination of old equipment and facilities. This can slightly modify the ranking of measures in terms of total gas savings.

Today, however the potential for Gazprom to reduce emissions using low compressor ratio transportation regimes is limited by the increasingly high capacity use of the pipeline system, from increasing Russian gas production and plans to dramatically increase gas imports from Central Asia. In this case, modern technologies of maintenance and repair are expected to lead to a 50% reduction of total potential GHG emissions (18-21 $MtCO_2e$), while replacement and modernisation of compressor stations could provide another 30% of the reductions (11 $MtCO_2e$).

Gazprom's set of options to mitigate GHG emissions is based on estimates of its Energy-Saving Programme for 2001-10 (Gazprom, 2001b) which provides "low" and "high" cost options. There are two types of low cost options:

- The first has no net (specific) costs, because energy savings is an ancillary benefit. The main objective of these options is to maintain and develop the transmission system's capacity. For example, the use of preventive internal pipeline diagnostics aims primarily to reduce the number of accidents and improve the reliability of the system. Maintenance of compressor units to ensure their optimal functioning also improves energy efficiency and reduces fugitive emissions. These options can be considered as "first order".

- The other options can be considered as "second order", as they have a net cost of energy savings. However, the capital investment outlays specific to energy savings is not significant, as for example the implementation of modern repair practices listed in the next section.

The major low cost GHG emission reduction options are concentrated at various stages of reconstruction (67% of energy-saving potential) and operations (18% of total).[107] This highlights the synergies between current needs of refurbishment of Russia's gas transmission system and the potential climate-driven investments. From 1990 to 2001, the bulk of these investments were postponed due largely to the barriers highlighted in Chapter 1.

During 2002-03, Gazprom achieved natural gas savings in the transmission system of 4.6 bcm (equivalent to almost 10% of Gazprom's own use in 2003). In 2004, this achievement was surpassed with savings of 3.5 bcm in that year alone. Gazprom also reported a very attractive return on its energy-saving projects – USD 90 million against costs of USD 25 million (Gazprom, 2004b).[108] Clearly, Gazprom has a large potential to replicate similar first and second order gas-saving projects.

In the near future, Gazprom may have to choose between investing in more expensive and difficult-to-develop fields within its portfolio or opening the market to independent gas producers, if it is unable to carry out its strategy of importing relatively cheap gas from Central Asia. It is, therefore, important for Gazprom to quicken the pace of

---

107. Potential energy savings during new construction stages, estimated at 15% of total, are high cost options.
108. If one were to calculate the value of saved gas at the export price of gas, benefits would be higher.

its gas-savings programmes – and Kyoto-related flexibility mechanisms could help Gazprom achieve more savings more quickly.

GHG emission mitigation through improved maintenance and operation practices

Gas-saving projects implemented by Gazprom during 2002-03 mainly focussed on optimising the transmission system. Gas savings attributed to these projects amounted to 30% of total saved gas over this period and 57% of savings in 2004.

According to Gazprom's Energy-Saving Programme (Gazprom, 2001b), the options at the operation stage are mostly of first and second order and include:

- Optimising technological regimes.

- Cleaning pipelines.

- Implementing modern technologies in maintenance and repair work, including preventive internal diagnosis along pipelines, adding new pipeline branches without blowdown (under pressure), leak detection and repair, valve replacement and gas recuperation before maintenance and repair.

Below we consider existing international experience as well as that of Gazprom in implementing energy-saving options.

Reducing blowdowns before maintenance and repair work

Blowdowns prior to maintenance and repair work are a significant source of methane emissions along transmission pipelines. They represent up to 60% of fugitive methane emissions in Russia's gas pipelines. Natural Gas STAR (2004a,b) suggests three main options to reduce blowdowns:

- **Pump-down techniques** consist in using fixed and/or portable compressors to lower gas-line pressure before maintenance. In both cases, reduction of the pressure in the affected segment is normally enough to allow the insertion of sleeves over the damaged area so repairs may begin. The pipeline pump-down technique permits the recovery of 50-90% of the gas typically vented. In-line compressors are used either alone or in a sequence with portable compressors (to obtain an additional 40% of gas recovery). Use of in-line compressors is usually cost-effective because it does not require additional equipment; *i.e.* there is an immediate payback in terms of marketable gas. Gazprom states in its 2004 Environmental Report, that Volgotransgaz reduced its CH_4 emissions by 23% compared to 2003 levels by using this technique (Gazprom, 2004b).

- **Ejectors** can recover gas from blowdowns. This technique requires the injection of natural gas into a lower pressure pipeline, using gas from a nearby higher-pressure gas pipeline.

- **The combustion option** consists in bleeding some of the high-pressure gas from in-service equipment into a low-pressure fuel gas system. By recovering a part of the volume for fuel, dual benefits of reduced fuel costs and decreased emissions can be obtained. In cases when blowdown gas cannot be recovered for fuel use, a flare installation (portable or fixed) can be used to combust gas (as a last resort).

Preventive
diagnosis and
maintenance
work

Routine maintenance is necessary to prevent potentially dangerous blowouts of CH_4 in the system from cracks and other pipeline deformities. In 2004, Gazprom reduced its CH_4 emissions by 0.15 bcm of CH_4, largely due to preventive diagnosis and maintenance work (Gazprom, 2004b).[109]

A "diagnostic pig" can be used to assess the integrity of the pipeline system. This is a balloon-like tool inserted into pipelines to clean the internal surfaces as well as identify any defective section. The expansion of the use of intelligent pigs and rehabilitation of gas pipes provide an opportunity to further reduce the number of accidents. The use of pigs along Gazprom's transmission pipelines intensified over the 1990s, and many potential pipeline incidents were prevented in this way. Since 1991, over 112 000 km of pipelines were inspected with diagnostic pigs (see Figure 18). The annual rate of inspection of about 20 000 km reached in 2003 is considered as optimal by Gazprom (Budzulyak, 2004). However, the use of pigs depends on specific pipeline parameters. For instance, the diameter of the pipeline must be sufficient. This tool also requires the installation of launch and reception stations. Currently, only 35% of Gazprom's transmission system is adapted for this type of equipment.[110] Gazprom is installing the necessary equipment during routine maintenance and repair operations. It is also using external methods of diagnosis to avoid limitations due to system design in implementing diagnostic pigs.[111]

Optimisation
of operational
regimes: a
Gazprom and
Ruhrgas project

Optimisation of operational regimes of transmission pipelines is another promising option to reduce energy consumption of the transmission system, and by extension contribute to the mitigation of CO_2 energy-related emissions too. The optimisation of operational regimes was implemented in a 1997-99 Gazprom and Ruhrgas project along the Uzhgorod Corridor of the Volgotransgaz transmission system (a pipeline system of 6 parallel lines, 4 500 km long, with 6 compressor stations and 133 compressor units).[112]

This project was implemented in the framework of Activities Implemented Jointly (AIJ) established by the UNFCCC's Berlin Mandate.[113] The optimisation work adjusted the use of individual compressor stations to minimise fuel gas consumption and hence reduce CO_2 emissions over the entire system of the Uzhgorod Corridor. The process involved offline optimisation using the SIMONE simulation and optimisation software[114] (adapted to Russian conditions), as well as optimisation of control measures on the transmission system itself.

---

109. The volume of CH_4 emissions reported in the Environmental Report (Gazprom,2004b) differs significantly from GHG emissions shown in Table 2.
110. About one third of pipelines need to be reconstructed to allow the use of "pigs" (Leontiev and Stureiko, 2003).
111. For more detail see *Energy Security of Russia* (2005).
112. Ruhrgas and Gazprom, 1997.
113. Activities Implemented Jointly consisted in using principles of JI without implementing real trans-border transfers of generated GHG emission reductions.
114. The SIMONE software package is implemented in more than 20 countries (over 200 installations) to manage dispatching of gas flows in transport system in on-line mode. It is compatible with various SCADA systems irrespective of the manufacturing country (Leontiev *et al.*, 2003a).

The baseline gas consumption of compressors along the Uzhgorod Corridor was estimated at about 3 bcm per year resulting in emissions of 6 $MtCO_2$. Through the overall optimisation of the transportation-operating system, total gas savings were estimated at 75 Mm^3 per year which corresponds to emission reductions of 150 000 tCO_2/y. These results were estimated by the innovative simulation method using the high modelling quality of gas pipeline operation in the SIMONE software.

The second phase of the project was due to begin in 2000 and be registered as a JI project. Gazprom and E.ON Ruhrgas developed a detailed computer model of the gas transmission network operated by Volgotransgaz (Leontiev *et al.*, 2003a). However, to date this project is still on standby. As indicated in Chapter 2, delays are often related to the lack of clear rules on national climate policy. This creates uncertainty for Gazprom over possible future constraints or benefits related to GHG emission reduction credits.

The second phase of the project encompasses a 10-stream pipeline system (Uzhgorod Corridor, Yamburg-Tula and Perm-Centre) covering 800 km and consisting of 7 500 km of pipes and 52 compressor stations with 232 compressor units.[115] The fuel gas savings in this phase are projected at 90 Mm^3/y, with reduced electricity consumption of 579 MWh and CO_2 reductions of 447 000 tCO_2/y. According to Ruhrgas, implementing this system over Gazprom's entire transmission system could lead to a reduction of 4-5 $MtCO_2$ per year or over 5% of total energy-related CO_2 emissions of Gazprom's gas transmission system.[116]

Refurbishment of compressor stations: the "double dividend" option

From 2001 to 2005, Gazprom planned to replace 551 compressor units by only 374 new more energy efficient units with higher unit power capacity (5.3 GW) as well as upgrading 437 other compressor units (5.1 GW).[117] This corresponded to replacements of 110 compressor units annually with 75 more efficient units and upgrading 87 other compressor units. According to the Gazprom Development Programme (2002-10), these measures could lead to improvements in the average efficiency of its gas-driven compressor units from 28.2% in 2000 to 31.4% in 2005 and to 33.4% in 2010, and to an increase of the average unit capacity of gas drives from 12.5 MW in 2002 to 14.5 MW in 2010 (see Table 17).[118]

Refurbishment needs of Gazprom's compressor stations to 2010

According to Gazprom's Energy-Saving Programme, gas savings at compressor stations should make up the largest share of total savings (see Table 16). The environmental benefits could be even higher if the substantial related reductions of CH_4 emissions were taken into account by Gazprom. The reduction of CO_2, in addition to gas losses (through reduced fugitive emissions), provides compressor refurbishment projects with a double dividend. This is increasingly attractive as domestic gas prices become more cost-reflective. However, according to Gazprom (2004b), upgrading and overhaul at compressor stations represented 35% of gas savings in 2002-03 (or

---

115. For more details, see E.ON Ruhrgas Web site www.eon-ruhrgas.com/englisch/umwelt/828.htm.
116. Point Carbon, 19 March 2004.
117. Upgrading can encompass the replacement of the drive engine, the installation of combined cycle and changeable flow-through (rotating) parts, supporting and sealing units, etc.
118. By 2005 the average energy efficiency of gas-driven compressor units should reach 31.4% (IBRC, 2003). However, no information is yet available on the results of this programme.

Table 17   Gazprom's outlook for improvements in its stock of gas-driven compressors

	2002	**2005**	**2010**
Total installed capacity of compressor stock (GW)	35.0	38.0	41.0
Average unit capacity (MW)	12.5	13.0	14.5
Average efficiency (%)	29.4%	31.4%	33.4%

Source: IBRC, 2003.

about 0.8 bcm/y) and only 21% in 2004 (or about 0.7 bcm/y). Based on the actual rate of gas savings due to replacement, modernisation and repair of compressors, and the objectives for gas savings announced by Gazprom in its Environmental Report (2004b) for the period 2004-06, total gas savings could be about 6 bcm from 2005 to 2010 (or about 1 bcm/y).

At the same time, assuming timely upgrades by Gazprom of the average efficiency of its gas-driven compressor units, in line with its Development Programme, total gas savings could be about 19 bcm in 2010 in comparison with its reported consumption level in 2000. These savings are mainly from increasing the average efficiency of its compressor units from 28.2% to 33.4%. Depending on the average efficiency attained in 2004, gas savings could be estimated between 6.6 to 9.6 bcm from 2005 to 2010. In addition to reductions of CO_2 emissions from gas combustion at compressor units, this would also reduce CH_4 leaks at least in the order of 1.2 bcm or about 20 $MtCO_2e$. This estimate is based on the assumption that Gazprom will continue replacing, upgrading and repairing its stock of compressors at the rates in its Development Programme. This could generate supplemental environmental benefits given that the replacement of old units can lead to a reduction of up to 80% of the CH_4 leaks from compressor stations (see below).

Reducing CO_2 emissions at compressor stations

The Canadian gas transmission system provides useful insights as it has a similar configuration of compressor stock in that the majority is gas-driven. Between 1990 and 2002 TransCanada (Canadian Natural Gas Pipeline Company) dramatically improved the efficiency of its compressor stations. In doing so it was able to limit the increase in CO_2 emissions to just 24% while increasing gas throughput and deliveries by 50% (see Box 6). TransCanada implemented the following GHG emission mitigation measures along its transmission system:

■ The use of the most energy-efficient engines available when adding capacity, retiring existing engines or purchasing new ones for pipeline-expansion projects.

■ The installation of variable frequency drives that match changes in throughput levels, or the use of larger diameter pipelines requiring less gas pumping units.[119]

■ The recuperation of waste heat from its compressors to power combined cycle gas turbines that generate electricity.[120]

---

119. Canadian transmission pipelines are large-diameter up to 1 219 mm or 48 inches (CEPA, 2003).
120. Recuperation of waste heat is implemented in Russia. In 2004, Volgogradtransgaz upgraded 5 compressor stations with heat-using facilities, reducing gas consumption by 40% (Gazprom, 2004a).

Box 6   TransCanada's gas transmission system

TransCanada's 37 696 km gas transmission network transports the majority of western Canada's gas production to growing markets in Canada and the United States. It transported 181 bcm of natural gas in 2003. The system is serviced by 110 compressor stations, including 258 compressor units with a total capacity of 3 400 MW. Between 1990 and 2002, the total length of TransCanada's system increased by 34%, and its compressor capacity by 78%.

In 2003, TransCanada's pipeline system's total CO_2 emissions amounted to 7.2 $MtCO_2e$, mainly due to compressors. In 2003, 92% of TransCanada's compressor power was derived from gas turbines, 2% from reciprocating engines and 6% from electric drives. The majority of the turbine engines are jet engines that can have thermal efficiencies of up to 39%. The average efficiency of TransCanada's network is in the order of 35%, as the older engines have efficiency ratings of only 25%.

Source: TransCanada, 2003.

The efficiency of compressor stations can also be increased by using computerised engine control systems to optimise fuel consumption of gas drives by essentially converting standard rich-burning engines into lean-burning engines. This control system can lead to a 15% reduction in gas consumption from engines at a cost of USD 1.2 million per unit (C^3 Views, 2005).

Reducing fugitive CH_4 emissions at compressor stations

It is important to prioritise the actions to be taken to reduce emissions given the number of possible sources of leaks and vents at compressor stations. This can be based on the Best Management Practices (BMP), reported by American natural gas pipeline companies in the framework of the Natural Gas STAR Programme, which includes the Method of Directed Inspection and Maintenance at compressor stations (DI&M) (Natural Gas STAR, 2003a).

In 1996, the Gazprom and US EPA project tested the applicability of the Natural Gas STAR DI&M at Russian compressor stations (Gazprom and US EPA, 1996). The study concluded that this method could be directly applied in Russia and could significantly optimise emission reduction efforts by contributing to the development of a comprehensive and economic methane monitoring/mitigation system. The main steps of the DI&M at compressor stations are highlighted in Table 18.

The replacement of leaky components is an alternative to replacing the whole compressor. Natural Gas STAR (Robinson *et al.*, 2002) recommends: (i) replacing wet seals with dry seals in centrifugal compressors, and (ii) reducing compressor blowdowns when taking them off-line.

Replacing wet seals with dry seals

Compressor seals are designed to prevent high-pressure gas from escaping. Wet seals use oil circulated under high pressure to form a barrier to prevent natural gas from escaping. Wet oil seals cause the most common form of leaks at centrifugal compressors. The CH_4 leaks mostly occur when the circulating oil is stripped of the gas absorbed at the high-pressure seal face.

Table 18   Natural Gas STAR Method of Directed Inspection and Maintenance (DI&M) at compressor stations

Directed inspection & maintenance at compressor stations		
Steps	**Operations**	**Techniques / Devices**

1	Conduct screening and measurements	Screening techniques:
		  Soap screening
		  Electronic screening
		  Toxic vapour analyser
		Devices quantifying emissions:
		  High-flow sampler
		  Rotameter

2	Evaluate results	Component regrouping according to leak rates
3	Prioritise and repair leaks	Selection of leaks to be repaired
4	Develop survey plan	Design future surveys according to a list of component classes and collected data

Source: Natural Gas STAR, 2003a.

The rate of leaks from wet oil seals increases with ageing. The Gazprom & Ruhrgas study showed that the seal oil systems at the older Kazym compressor station (built in 1971-77) were responsible for half of the increased fugitive emissions compared to those of the newer Upper Kazym station (built in 1983-97). Leaks at the older Kazym station amounted to 27 470 m^3/MW per year or nearly 80% more than the seals oil system of the newer Upper Kazym compressor station.

Dry seals, which use high-pressure gas to seal the compressor, emit from 6 to 33 times less methane and have lower power requirements. This improves compressor and pipeline operating efficiencies and performances, enhancing reliability and requiring less maintenance (see Table 19).

While some compressor designs prohibit installation of dry seals, the experience of Natural Gas STAR participants indicates, that where installation is possible, conversion from wet to dry seals makes economic sense, in most cases.[121] Given lower natural gas prices, as well as equipment and labour costs, it is unclear if this process is economic in Russia.

Table 19   Comparison of wet and dry oil seals

	Gas leakage rates, m³/year	Electricity consumption, kWh	Reliability	Annual O&M costs, USD/year
Dry seals	10	5	Higher	6 000 - 10 000
Wet seals	68-340	50-100	Lower	> 100 000

Source: Hanle, 2003.

---

121. In the United States, with natural gas prices of USD 106/thousand m³, the pay-back period is about 14 months with costs of around USD 135 000/year (Natural Gas STAR, 2003b)

Reducing CH_4 emissions from compressors taken off-line

The US EPA Natural Gas STAR partners report that considerable reductions of CH_4 emissions at centrifugal compressor stations can be obtained by changing operating practices using two alternative options (Natural Gas STAR, 2004a):

■ Keeping compressors pressurised, thus avoiding a blowdown when the compressors are off-line for operational reasons. This option does not imply capital costs and has an immediate pay-back.

■ Using normally vented gas as fuel by connecting the blowdown vent lines to the fuel gas system while the compressor is off-line.[122] However, this option is effective only where there is sufficient fuel demand to consume the otherwise vented gas (*i.e.* 40 m³/hour).

Role of carbon finance

Given the number of compressors to be upgraded or replaced by Gazprom annually, it is interesting to consider the possible impact of carbon finance on its technological choices and the rate of annual refurbishment. A generic project that replaces old compressor units (gas turbines and compressors) by more efficient ones is examined (see Table 20).[123]

The existing 25 MW unit (gas turbine) has a 25.6% energy efficiency corresponding to the energy efficiency of Gazprom's gas turbines to be replaced in its Energy-Saving Programme (Gazprom, 2001b).

Gazprom has stated its preference for nationally produced gas turbines and compressors. Currently, 86% of Gazprom suppliers of equipment for compressor stations are Russian companies. The remaining 14% is supplied by foreign companies.

The Russian new-generation drives, which are to be installed by Gazprom according to its Energy-Saving Programme have efficiency ratings comparable to the best

Table 20

Assumptions used to evaluate compressor replacement projects

Assumptions	Energy efficiency, %	Investment, USD/MW
Actual compressor unit	25.6%	–
New compressor unit:		
Foreign unit (higher efficiency)	38.0%	1 000 000
Russian unit (higher efficiency)	38.0%	500 000
Russian unit (business-as-usual)	35.0%	175 000
Domestic natural gas price, USD/thousand m³	40	
Export natural gas price, USD/thousand m³	260	

Source: IEA estimates based on discussions with experts and Gazprom information.

---

122. At natural gas prices of USD 106/thousand m³ the pay-back time is estimated at 4 months.
123. Either drive or compressor can be replaced. However, according to experts, both are generally replaced.

internationally-available technologies (BAT) (see Annex 4).[124] The new Russian 16 MW turbines are as efficient as the BAT (about 36%) but their 25 MW units are for the most part less efficient than the BAT.[125]

Three options are considered:

- Replacement by a new foreign-made 25 MW compressor unit with a 38% efficiency, at a cost of USD 1 million/MW.

- Replacement by a new Russian-made 25 MW unit with the same 38% efficiency, at a cost of USD 0.5 million/MW, and

- Replacement by a new Russian 25 MW unit with 35% efficiency, at a considerably lower cost of USD 0.175 million/MW.[126]

In this analysis, the economic benefit from refurbishment projects is assumed to be based primarily on increased gas sales over a 20-year project life given growing domestic and export gas demand.

Figure 20 presents the IRR for each of the three options using current and future domestic gas prices of USD 40/thousand m^3 and USD 60/thousand m^3, respectively, as well as an export gas price at the begining of 2005 of USD 170/thousand m^3 to value the saved gas. Experts can argue that the gas saved should be valued at the cost of production. The current and future domestic price is used as a benchmark of the increasing long-term marginal cost of Russian gas. The export price is a proxy of the indicator that the foreign investor could use to evaluate economics of compressor replacement projects in Russia if a joint venture with Gazprom could be established for this purpose.

Figure 20 shows that Gazprom would prefer the low cost business-as-usual compressor unit in its refurbishment programme, and that it has no incentive to choose more efficient and more costly units or refurbish its system at a faster pace than necessary in order to maintain capacity and stability. This figure also demonstrates that the price of gas is a key factor behind Gazprom's technological choice. The use of foreign BAT becomes attractive (IRR>30%) at current export prices of USD 260/thousand m^3 on European markets if these export markets can be reached.

In addition to gas savings, the project generates significant environmental benefits in terms of GHG emissions reductions that could be valued on the carbon market and add to the economic benefits of compressor replacement projects:

- The 38% efficiency units both Russian and foreign reduce gas consumption by 32% and CO_2 emissions by 54 000 tCO_2 per year.

---

124. Direct comparison is difficult given units have different power ratings and other parameters are not available.
125. The units with capacities higher than 25 MW are available only among the foreign BAT.
126. This is consistent with costs of new compressors in the Energy-Saving Programme of Gazprom (2001b).

Figure 20   IRR for compressor unit replacement investments at domestic and export gas prices (business-as-usual)

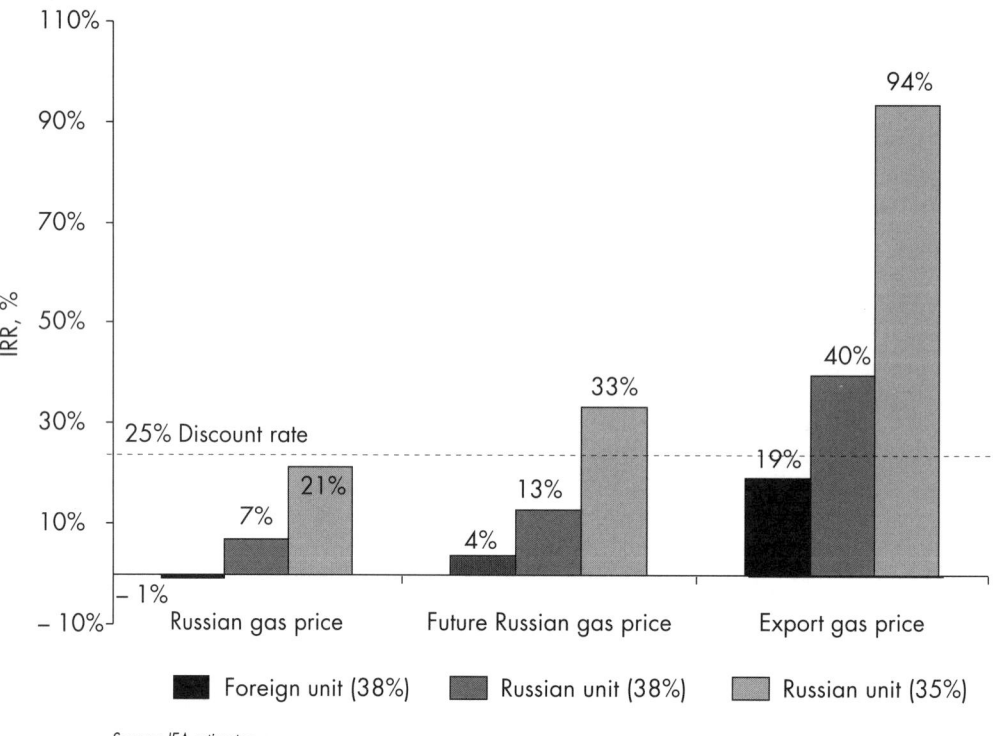

Source: IEA estimates.

- The 35% efficiency unit reduces gas consumption by 27% and CO_2 emissions by 49 000 tCO_2 per year.

- In addition to CO_2 emission reductions due to the more efficient drive, the refurbishment also leads to the replacement of old components of compressor units and reduces fugitive CH_4 emissions, by an estimated 28 000 tCO_2e per year.[127]

However, the possible contribution of carbon finance to project economics depends on the amount of GHG emission reductions that will be considered additional. To demonstrate additionality, an appropriate baseline needs to be determined to reflect the most representative and conservative business-as-usual alternative to the project (see Chapter 2). It is assumed that the investor could generate the ERUs during a 10 year period, and that during this period the baseline will remain constant.[128]

We assume that replacement of existing compressor units in order to increase the average efficiency of gas driven compressors units from 28.2% (in 2002) to 34.1% in

---

127. This assumes the project could reduce CH_4 emission levels from the high to medium value in Table 15.
128. The "crediting period" is the time when annual GHG emission reductions obtained by JI or CDM projects are accumulated (7-10 years for most CDM projects). This period reflects the time when baseline conditions remain stable and reflect real and sustainable reductions of GHG emissions by the project. It should be reviewed for subsequent periods given the possible evolution of technology and economic situation.

2010 is the appropriate baseline. It reflects the current and projected compressor unit replacement measures in Gazprom's Development Programme (2002-10). In this case, for our example, the baseline scenario is the replacement of the 25% efficiency unit by the Russian business-as-usual 35% unit. Only a project that would refurbish using compressor units with 38% efficiency is considered as additional. In this case, if the investor chooses the more efficient alternative, the supplemental environmental benefits (of about 10 000 tCO_2/year) generated by the increase of energy efficiency of the unit from 35 to 38% would be considered as additional.[129]

At a CO_2 price of USD 7/tCO_2, the current low-bound price for forward JI transactions, the carbon revenue represents over 6% of the total project revenues assuming the gas savings are sold on the domestic market. Using the upper-bound price of USD 14/tCO_2, carbon revenues represent over 10% of annual project revenues.[130]

Implementation of the BAT leads to significant additional GHG emission reductions given the 20-year lifespan of compressor units. A simplified IEA calculation estimates possible additional gains in terms of annual GHG emission reductions at over 1 $MtCO_2$e for every 100 compressor units replaced, which is Gazprom's reported annual rate of replacement. However, the impact of carbon finance on the economics of the project is marginal and cannot be a driver for the technological choice at any of the assumed CO_2 price levels. With our highest assumed CO_2 price of USD 26/tCO_2 (EU ETS price as of end-2005), carbon finance could add less than 2% to the IRR of the project using a Russian higher-efficiency unit bringing it to 9%.

The above calculation shows that the carbon component of these capital-intensive projects will not be the main driver of investment decisions given, in particular, the current level of domestic gas prices. The progressive increase of Russian domestic gas prices would play a more important role here as an incentive for choosing the BAT. This confirms that carbon finance could be more effectively used in other sectors where it can provide a more substantial impact on investment decisions and help to overcome barriers for energy-efficiency improvements.

At the same time, the supplemental potential of energy savings may be implemented by Gazprom in the framework of JI, using new foreign practices and technologies of maintenance and repair. These measures are much less capital-intensive than the refurbishment of compressor stations and could generate *additional* GHG emission reductions to what could otherwise be obtained by Gazprom in a business-as-usual scenario. The implementation of such projects could be enhanced by long-standing export-import relations and partnerships between Gazprom and gas companies of EU countries, and more recent partnerships with Canadian and Japanese firms. Partnerships to enhance the transmission capacity of Gazprom's domestic infrastructure could become just as important as planned expansion of export routes in the coming years,

---

129. Only the additional CO_2 emission reductions are considered given that the CH_4 emission reductions could otherwise be obtained by a business-as-usual project.
130. The purchase price of CO_2 depends on the modality of risk sharing between buyer and seller of emission reduction units. This depends also on the stability of the Kyoto-related framework of the host country.

if Gazprom's domestic infrastructure is not refurbished quickly to match increasing domestic and export needs. In addition, with the era of *cheap* gas coming to an end, Gazprom should become increasingly interested in maintenance and refurbishment of its transmission system to reduce its own gas use and losses.

Investors should be aware of these wide-ranging factors involved in Gazprom's decision making process, and by the questions surrounding the use of Kyoto-related mechanisms. In terms of investments in the gas transmission system, Kyoto flexibility mechanisms could play a marginal role in stimulating a quicker rate of refurbishment and also stimulate the use of BAT and practices. This could have a significant impact on the reliability of Russia's transmission system – enhancing its throughput capacity and ensuring a growing number of domestic Russian gas producers, thereby promoting competition in Russia's upstream gas sector along with substantial long-term environmental benefits.

IV. REDUCING METHANE EMISSIONS FROM RUSSIA'S GAS DISTRIBUTION NETWORK

With 575 000 kilometers of high, medium and low-pressure distribution pipelines, Russia has the world's second largest natural gas distribution system behind that of the United States supplying about 380 bcm to Russian customers in 2004 (IEA preliminary data). In contrast to the many studies and international projects already undertaken on Gazprom's transmission system, little attention has been focussed to date on the potential reduction of GHG emissions within Russia's gas distribution network. Consequently, there is very little official or publicly available information on the state of this system (IEA, 2002).

During the latter part of the 1990s, gas distribution companies were privatised and became independent. Severe financial problems — principally resulting from non-payments by consumers — drove many distribution companies into insolvency. Since 1999, Gazprom has been increasing its control over this sector and in 2004 had interest in about 75% of gas distribution facilities (see Chapter 1). As in other parts of Russia's energy sector, low retail end-use tariffs and the lack of meters severely affected the economics of projects to maintain or upgrade the distribution network.[131] Most of the potential gas savings and related GHG reductions would be a by-product of commercially viable distribution if end-use retail tariffs were fully cost-reflective.

Recent studies and pilot projects in certain Russian regions and cities have shown that the potential methane emission reductions from ageing and poorly maintained distribution networks could be even larger than in the transmission system. However, given the nature of the distribution sector, investment projects to tap this potential will be small and dispersed over hundreds of municipal systems in comparison to investment projects in the gas transmission sector. Transaction and monitoring costs could render these projects unattractive. The limited financial and technical capacities of small local gas distribution organisations are another barrier. Kyoto flexibility mechanisms, or the Green Investment Scheme could provide a useful and timely lever to overcome investment barriers in this sector if they can provide the possibility to bundle similar small projects together. This would reduce transaction costs and thereby raise overall project returns.

This chapter examines the main components and current state of Russia's gas distribution sub-sector.[132] It attempts to provide rough estimates of the level of GHG emissions based on the current state of Russia's gas distribution facilities and on the results of a limited number of recent regional measurement programmes.

---

131. Retail prices were regulated by regional and federal Energy Commissions, now the Federal Tariff Service.
132. The ownership structure of this sector is described in Chapter 1.

International comparisons are made, mainly drawing on North American experience over the 1990s estimating CH_4 leaks along gas distribution systems. Finally, a description is given of available measures to reduce leaks in the gas distribution system and an assessment of the potential role of carbon finance in enhancing the scope of energy-saving investments in Russia, given current structural and market barriers.

CURRENT STATE OF RUSSIA'S GAS DISTRIBUTION SYSTEM

Main components of the gas distribution system

Gas distribution systems around the world contain several principal components: the gate stations (or city gate), the downstream pressure reduction stations, the underground high-pressure mains, medium and low-pressure pipelines, the pressure regulators and meters and the service systems bringing gas to the consumers (see Figure 21).

Gate stations are located at transfer points where natural gas is delivered from transmission pipelines into the high-pressure lines of local distribution companies. The gate station typically contains metering runs and pressure regulators that reduce the pressure of the dedicated high-pressure pipeline branches (connecting transmission pipelines to the gate stations) to usually less than 20 atm (atmosphere). Other surface facilities within the distribution system include downstream pressure regulators further reducing gas pressure so that gas can be delivered safely to consumers by means of low-pressure mains, usually located underground.

Figure 21

The general schematic of natural gas distribution networks

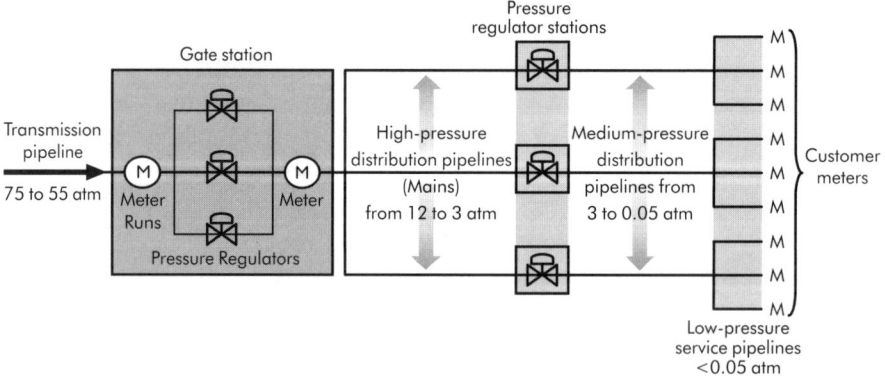

Source: based on Natural Gas STAR (2003c) and Gosgorteknadzor (2003).

In Russia, the gate stations reduce pressure from 55 atm to about 12 atm to supply gas into the high-pressure distribution pipelines. The pressure must match the regimes of two other categories of high-pressure pipelines (from 6 atm to 3 atm), medium pressure pipelines (from 3 atm to 0.05 atm), and low-pressure service pipelines (less than 0.05 atm) (Gosgorteknadzor, 2003). The extensive range of pressure regimes and diameters of service pipelines in Russia compared to distribution systems in the

United States and Canada limits the use of the emission factors developed by these two countries to assess Russian emissions. It also clearly points to the critical need to develop specific emission factors for the Russian system.

Gas distribution pipelines and facilities

Russian pipelines are predominantly made of steel, although plastic is being used increasingly and accounts for a quarter of new underground pipelines installed in 2005 (Gosgorteknadzor, 2005). Plastic pipelines have the advantage over steel pipes in that they do not corrode and have an operational lifespan twice that of steel pipelines. However installation of plastic pipelines is a more expensive undertaking.

Table 21 shows that as of 2004, over half of Russia's natural gas distribution network is relatively new (less than 10 years old).[133] However, an increasing part of the remainder is close to or already past its normative lifespan of 40 years. In 2004, close to 26 000 km or 7% of Russia's stock of underground distribution pipelines was at or over its normative lifespan of 40 years (Gosgorteknadzor, 2005). According to Vasiliev (2005), the share of underground pipelines passed their normative lifespan is doubling every 5 years.

The ageing of the distribution pipeline system and current poor maintenance and repair practices increase the risk of corrosion. Normally, preventive diagnostic and maintenance procedures can extend the operational life of the network. In Russia, however, only one quarter of the underground pipelines over 40 years old were monitored and repaired in 2004 (Gosgorteknadzor, 2005).

Russia's gas distribution system includes 3 800 gate stations, essentially owned by Gazprom, and 113-143 thousand regulating stations. There is little information available about their current technical and operational condition, partly due to the rapid change in ownership over recent years. What is known is that over 10% of Russia's regulating stations have passed their normal operational lifespan of 20 years (Gosgorteknadzor, 2005).

Table 21   Expansion of the gas distribution system from 1995 to 2004

	1995	1996	1997	1998	1999	2000	2001	2002	2003	2004
Length, 1 000 km	278	311	340	371	393	424	452	497	514	575
New pipelines	n.a.	33	62	92	115	146	173	218	235	297
% of new pipelines	–	11%	18%	25%	29%	34%	38%	44%	45%	52%

Note: The huge increase in length of pipelines in 2003-04 may be explained by more comprehensive information available after the 2004 inventory by Gazprom of its distribution facilities and its growing control over regional distribution companies since 1999.

Sources: Vasiliev, 2005; Gazprom, 2004; IEA estimates.

---

133. The rapid expansion of the gas distribution network (by about 7% per year) has recently been boosted by Gazprom's regional gasification programme (see Chapter 1).

Table 22   Accidents along Russia's gas distribution network from 2003 to July 2005

Type of accident	2003	2004	July 2005
Mechanical damage of underground pipelines	15	15	8
Explosion during equipment startup	4	12	2
Mechanical damage of pipelines by automobiles	1	6	1
Pipeline corrosion	1	5	2
Breaking of steel connections	2	4	3
Damage due to natural forces	4	1	1
Excess pressure after gas regulating stations	2	5	1
Other	4	5	3
Total	33	53	21

Sources: Gosgorteknadzor, 2005.

Accidents in the gas distribution system

Table 22 presents official Russian statistics on accidents in its gas distribution system reported by Gosgorteknadzor (2005). It provides useful insight on the system's safety and reliability.

From 2003 to the first half of 2005, the number of accidents in Russia's distribution system remained relatively stable. Although the data covers a very short timeframe, accidents relating to corrosion, the breaking of steel connectors and excess pressure after passing gas-regulating stations increased from 15 to 30% of total accidents. According to Gosgorteknadzor, this is largely due to low-quality maintenance, poor diagnosis and the subsequent lack of refurbishment of gas distribution facilities and pipelines.

Technical and commercial losses in Russia's gas distribution system

Existing estimates of gas losses in Russia's gas distribution sector are highly uncertain as there is limited information about the current state of equipment, the number of potentially leaking components and specific emission factors. That being said, Gazprom announced in its 2004 Annual Report the completion of a comprehensive inventory of its gas distribution facilities determining the current state of equipment and its reconstruction and maintenance work. However no further detailed information was provided.

Russian experts often note the lack of meters at end-use points and accurate information and data on exact levels of natural gas consumption by end users. The significant difference between the volume of natural gas which is supplied and that which is consumed is what Russians call the "imbalance". Russian experts include "commercial losses" into this "imbalance", a term used to describe "non-sanctioned" consumption (or theft). This could also be due to metering errors, unaccounted sales, or faulty book-keeping. Some experts argue that the share of these "non-sanctioned" losses could account for up to 70% of the "imbalance". The remaining 30% of the "imbalance" is attributed to operational fugitive emissions (leaks in seals, repairs) as well as those due to accidents along the gas distribution network.

Some Gazprom experts argue that the bulk of the "imbalance" is due to the lack of meters at consumption points as only 30% of households are equipped with meters (Zhilin, 2004). Gazprom officials also note that most installed meters are old and have a relatively high rate of error of 5% compared to modern meters that have a margin of error of 0.25% (Gazprom, 2001a; Remizov *et al*, 2000). Furthermore, the accuracy of measurements can be greatly reduced by the quality of the gas itself, the presence of water in the gas, condensate, oil and impurities. Significant differences can also occur given that most meters are not equipped with adapters for pressure and temperature. Each 10°C below normal levels of temperature leads to a 3.5% lower evaluation of gas consumed, if the meter is not adapted to correct for significant temperature changes (Zolotarevskiy and Osipov, 2004). Given average winter temperatures in Russia, this could account for a large share of the "imbalance".

GHG EMISSIONS IN THE GAS DISTRIBUTION NETWORK

To date there are no verifiable estimates of the level of gas losses during normal operations or due to leaks and accidents in Russia. This is mainly due to the lack of meters at end-use points, given that residential and municipal gas consumption is billed based on norms and not on actual gas consumed.

Gazpromregiongaz and Promgaz[134] estimate the potential energy savings in the gas distribution sector between 5-10% of the volume of distributed gas, including technological savings and reductions of commercial losses (Karasevich and Terekhov, 2004). Based on an estimated volume of 165 bcm of natural gas distributed through the medium and low-pressure network in 2004, this corresponds to between 8 and 17 bcm.[135] If we assume, as explained above, that one third of these losses are attributed to operational leaks, then 3 to 6 bcm of gas per year could be saved.

According to Gazprom (2001a), the level of normative (officially allowed) losses is between 0.9 and 2.2% of the total volume of distributed gas. Losses that Gazprom can charge consumers as "allowable losses" is even lower, at 0.5-0.6% (FTS of Russia, 2005). In 2004, therefore, sanctioned losses were no less than between 1.5 and 3.7 bcm of natural gas, equivalent to emissions of about 23-55 $MtCO_2e$.

Studies on GHG emissions in gas distribution in the United States and Canada

A wealth of detailed information on emissions is available from various studies of the United States and Canadian gas distribution systems undertaken in 1996 and 1998. Their results are extensively used in GHG estimations and reporting in North America. Estimates of GHG emissions are derived using specific emission factors by type of technical component and a comprehensive inventory of existing facilities. However, current methodologies for GHG inventories in the gas distribution sector, developed by the US EPA and the Gas Technology Institute in Canada are region-specific. Despite this, they provide useful insight on possible major sources of GHG emissions within Russia's gas distribution system.

---

134. Promgaz is Gazprom's planning and design centre for gas distribution and utilisation.
135. The volume of gas distributed through low and medium-pressure pipelines needs to be estimated given that no statistics are available on the gas distributed to consumers by category of different pressure regimes. The estimate of 165 bcm for gas distributed through low and medium-pressure pipelines in Russia in 2004 is based on IEA preliminary data of 420 bcm less the volume supplied through high-pressure distribution pipelines to heat and power producers, big industrial consumers, and the volume of gas consumed by transmission pipelines. This represents about 40% of total distributed gas.

GHG emissions in the gas distribution network are related to the losses of distributed gas during normal operations, repairs and preventive leaks (venting) and accidents. CO_2, CH_4 and nitrous oxide (N_2O) are the main GHG emissions produced directly by gas distribution systems during the operation and maintenance of pipelines. CH_4 is by far the most important GHG emitted by the gas distribution system, representing 90% of emissions. CO_2 emissions represent only 3% and are mainly from energy combustion of heating equipment installed in gas distribution pipelines. The volume of emissions from this sector depends significantly on the materials used in the mains, the age of the equipment, the pressure at different points along the system, and the maintenance records.

The 1992 measurement programme by the US EPA and GRI (1996a) showed that methane emissions from the distribution system are largely due to underground pipeline leaks from seals and fittings (54%), to leaks at metering and pressure-regulating stations (35%) and to leaks from meters at end-use points (8%).[136] This study estimated total losses of gas in the United States' distribution system at 0.5-0.8% of distributed gas. This is two to three times less than the normative rates of gas losses of 0.9-2.2% in Russia's distribution system according to Gazprom. Russia's low-end value is higher than the high end value listed for the United States. Furthermore, the United States' rates are based on studies undertaken in the early 1990s. As such, these estimates are likely over-estimations of current emission rates in the United States, given the cost-cutting incentives for private companies and high retail gas prices since then. In this respect, emissions from Russia's gas distribution sector should provide attractive opportunities for investors to reduce natural gas losses and GHG emissions.

Fugitive methane emissions at underground gas distribution mains and service pipelines

Underground pipeline leaks represent the largest part of CH_4 emissions in the United States and Canada. These are caused by corrosion, material defects, joint and fitting defects or failures (US EPA and GRI, 1996b). The extent of pipeline leaks depends on the composition of pipelines and the pressure of natural gas within the system. As shown in Table 23, mains constructed of cast iron are much more susceptible to leaks than pipelines constructed from protected steel or plastic. Protected steel mains have the lowest leakage rates. In terms of service pipelines, plastic has the lowest leakage rate. In general, leaks at the service end of the system (given lower pressures) are much lower than at mains.

Methane emissions at distribution mains and service pipelines can also be due to releases of natural gas during maintenance and repair work. They need to be purged when new lines are installed or older lines are replaced or repaired. The volume of vented natural gas by kilometer depends on operational techniques. For instance, companies such as Enbridge in Canada release the natural gas from pipelines which are to be put out of service into adjacent lines prior to purging, providing a much lower emission rate.

---

136. The US EPA and GRI (1996b) estimates do not include 18% of natural gas leaks from underground distribution pipelines that are oxidised in the soil. This information is not available for Russia.

Table 23   Rates of CH_4 leaks along underground distribution pipelines in the United States, 1992

Pipeline use	Pipeline material	Emission factors
Main pipelines		m³/km
	Cast iron	4 250.0
	Unprotected steel	1 690.0
	Protected steel	54.5
	Plastic	70.2
Service pipelines		m³/service pipelines
	Unprotected steel	48.7
	Protected steel	5.0
	Plastic	0.3
	Copper	7.3

Source: US EPA and GRI, 1996b.

Leaks at gate stations, regulating and metering facilities

An additional source of leakage in the distribution system is at the gate station and other surface facilities where, over time, various components can develop leaks due to temperature fluctuations, pressure, corrosion and weather. Studies have shown that the largest leaks are generally located at various pressure relief valves. Studies conducted in 1994 and 1998 in the United States found that methane released during normal operations of pneumatic devices (controllers) accounted for over 95% of total site emissions (Natural Gas STAR, 2003c).

Gate stations and surface facilities vary significantly in size and inlet pressure capacity depending on the scale and complexity of the distribution system. In general, the higher the inlet pressure, the larger the gate/regulating station and the greater the number of equipment components that may develop leaks. Table 24 presents the emission factors determined by the US EPA's 1992 measurement programme metering and pressure regulating stations in the United States.

Studies have shown that at similar levels of inlet pressure, emission factors are also higher for stations with both pressure regulators and meters encompassing larger numbers of potential leaking components. Regulating stations located in vaults have significantly lower emission rates than above ground stations of similar inlet pressure.[137] The US EPA emission factors provide a typical ranking of emission sources and indicate potential priority areas in measurement and maintenance programmes.

In Canada, pipelines also represent the main source of methane emissions (see Table 25) as reported by Enbridge, the largest gas distribution company. Fugitive emissions of pipelines include emissions from underground valves and fittings and emissions through the pipe wall itself due to deterioration, corrosion and defects (Enbridge, 2003). In 1990, fugitive emissions represented about 0.2% of gas sales of Enbridge and this rate was reduced to 0.1% of sales in 2001.

---

137. These facilities do not have atmospheric-bleed regulators for safety reasons (US EPA and GRI, 1996c).

Table 24 Emission factors for gas distribution and pressure-regulating stations in the United States, 1992

Station type	Inlet pressure, atm	Emission factor 1 000 m³/year
Metering & pressure-regulating	> 20.4	160 560
Metering & pressure-regulating	6.8 - 20.4	85 370
Metering & pressure-regulating	< 6.8	3 839
Regulating	> 20.4	144 575
Regulating -Vault	> 20.4	1 160
Regulating	6.8 - 20.4	36 166
Regulating-Vault	6.8 - 20.4	179
Regulating	2.7 - 6.8	893
Regulating-Vault	2.7 - 6.8	89
Regulating-Vault	< 2.7	89

Source: US EPA and GRI, 1996c.

Table 25 Methane emission reductions of a Canadian gas distribution company, 1990 vs 2001

Fugitive emissions	1990		2001		2001/1990	Structure	
	Mm³	ktCO₂e	Mm³	ktCO₂e		1990	2001
Pipelines	11.9	179.5	8.4	126.5	- 30%	62%	52%
Equipment	5.5	82.3	6.8	103.1	25%	28%	42%
Process venting	1.3	20.2	0.3	4.1	- 80%	7%	2%
Third-party damages	0.6	9.0	0.7	10.2	14%	3%	4%
Total	**19.3**	**291.0**	**16.2**	**244.0**	-	**100%**	**100%**

Source: Enbridge, 2003.

Estimates of methane emissions in Russia's gas distribution system

To estimate the volume of total emissions from the entire distribution system, a broad and consistent set of information is needed about activity factors or the number of facilities and equipment for each particular range of emission factors. This type of information is available in the United States and Canada. Based on this data, specific emission factors were developed over the 1990s. In Russia, this type of detailed data is not yet available. Even if the North American emission factors were to be used in the interim – as approximations – while detailed data is being collected, they do not include emission factors for above-ground pipelines. Given that about 30% of Russia's gas distribution pipeline system is above-ground, more measurement programmes are needed. Furthermore, North American emission factors can not be used directly in Russia given different design and maintenance practices.

Estimates used in this study are based on available Russian regional measurements, Gazprom standards and norms, as well as the estimates of various Russian experts. These estimates encompass a high level of uncertainty given the small scale of samples of recent Russian measurement programmes, the lack of information on the whole system and the possibility of wide differences in complexities of regional networks, the age of existing facilities and maintenance practices.

Rosgazifikatsia, in co-ordination with Gazpromregiongaz and the support of foreign partners, is planning extensive on-site measurement work to develop specific emission factors and measurement methodologies for Russia's gas distribution sector. This would enable Russian gas distribution companies to put in place a good quality GHG inventory and to evaluate further potential GHG emission reductions. This would allow Russian companies to value GHG emission reductions in the framework of domestic and international climate policy instruments. This work has been planned since 2005 after a first pilot project was undertaken in Kaliningrad in 2003-04. At the time of completion of this book, results from regional measurement exercises in Kursk and Tver were available.

In 2003-04, a leak detection and repair programme was undertaken on the gas distribution network of the municipality of Kaliningrad. This project was supported by the Russian Regional Environmental Centre (RREC) and by the pilot project of the UK's Global Opportunities Fund (see Box 7). First-order low-cost refurbishment measures led to emission reductions of 20% at a cost of USD 25 000 or some USD 0.6/tCO$_2$e. Despite the success in Kaliningrad, Russian experts argue that its gas distribution system is not representative given its small size and isolation from the rest of the network. However, given the limited information available on leaks along the Russian distribution system, extrapolation of these results provides some preliminary insights (see Table 26).

Specific emission factors were calculated for the various components of Kaliningrad's gas distribution system using available data on the length of pipelines, the number of pressure-regulating stations and end-use equipment against the results of the methane emission estimates from the pilot project. These emission factors can be applied to the whole Russian gas distribution system, keeping in mind the high level of uncertainty of this extrapolation exercise.

Table 26   Estimates of methane emissions in Kaliningrad and extrapolation across Russia

Emission sources (estimated on-site)	Kaliningrad network			Extrapolation to national network		
	Units	CH$_4$ emissions, ktCO$_2$e/y	Emission factor, ktCO$_2$e/unit/y	Units	CH$_4$ emissions	
					bcm	MtCO$_2$e/y
Pipelines, km*	750	105	0.1400	276 000	2.6	39
Pressure-regulating stations	72	49	0.6806	142 500	6.5	97
End-use equipment	200 000	56	0.0003	44 000 000	0.8	12
Total	–	**210**	–	–	**9.9**	**148**

* For pipeline emissions, extrapolation is based on the length of pipelines more than 10 years old.

Sources: Kuraev and Safonov, 2005; RREC, 2004; IEA calculations.

Box 7   The Kaliningrad pilot project to provide "Support for the Russian gas industry to participate in Kyoto mechanisms"

This 2003-04 pilot project was led by the Russian Regional Environmental Centre (RREC), with financial support from the UK Global Opportunities Fund (GoF) and the British Embassy in Moscow. The main objective was to assess the effectiveness of reducing fugitive emissions in the gas distribution sector through low-cost replacement of tubing at major pressure points with a special material (teflon was used in Kaliningrad). The pilot project aimed at developing a specific methodology for estimating fugitive methane emissions in the municipal gas sector and demonstrating possible gas savings and environmental benefits.

A rough estimation was made of annual methane emissions for the city of Kaliningrad of about 200 ktCO$_2$e. The low cost (first-order) refurbishment of the network reduced this by a fifth or 40 000 tCO$_2$e at a total cost of some USD 25 000. This was found to be extremely attractive, having a pay-back time of less than 4 months assuming domestic natural gas prices of USD 35/thousand m^3.

The main outcomes of this project were:

■ Estimates of fugitive methane emissions from the gas distribution sector in Kaliningrad City.

■ Recommendations on the methodology for Rosgazifikatsia on fugitive methane emission detection for the municipal gas distribution systems.

■ Recommendations for the development of national methodological guidelines for a company-level fugitive methane emission inventory.

■ Demonstration of the technical and financial feasibility of such leak detection and repair programmes, including low-cost options to reduce emissions.

Sources: GoF, 2005; Kuraev and Safonov, 2005; Safonov, 2004.

This highly uncertain extrapolation exercise estimates total methane emissions from Russia's gas distribution network in the order of 10 bcm per year or 148 MtCO$_2$e. This represents roughly 6% of the volume of natural gas distributed through Russia's medium and low-pressure lines every year (165 bcm).[138] This is in line with the earlier Gazpromregiongaz and Promgaz estimates of potential gas savings in this sector excluding the "imbalance" factor.

A new measurement programme was implemented from August to November 2005 in the Kursk region by Centergazservice-opt and the Kursk regional gas distribution company Kurskgaz in the framework of a JI project proposal (Kursk JI PDD, 2005).

---

138. Losses reported in the city of Almaty in Kazakhstan are in the same order of magnitude at 4.9% of total distributed gas for 2000-02 (PNNL, 2004). Older and denser distribution networks reported losses between 7.5% and 9.2%. Regions in Kazakhstan undertaking timely refurbishment report losses of less than 1%.

This programme included an inspection of 71 regulating stations and detected that more than one third of inspected valves were leaking an estimated total of 1.1 Mm^3/y.

However given that only about 4% of valves operated by Kurskgaz were inspected, these results can not be considered as a representative sample. The project developers estimated gas losses for the entire number of Kurskgaz facilities (valves and flanges) at 21.4 Mm^3/y or CH_4 emissions of 322 $ktCO_2e/y$. According to Russian experts, this corresponds to about 1% of gas distributed by Kurskgaz. In 2006, another measurement project of 70 gas distribution facilities was conducted in Tver with the participation of Japanese companies and the rate of leaks were estimated at 1.5% of distributed gas.[139]

Extrapolation of Kursk results over the entire number of regulating stations in Russia, indicates that fugitive emissions could be in the order of 2.2 bcm or 30 $MtCO_2e$ per year. This value is about one third of the estimates based on the Kaliningrad emission rates and Gazpromregiongaz and Promgaz estimates (see Table 27). However, Kursk results do not take into account other potential sources of leaks. These could account for 50% of total losses according to North American data. Thus, total losses in Russian distribution pipelines could be double at about 4.4 bcm per year if one includes these other potential leaks. This represents about 2.6% of gas distributed through medium and low-pressure pipelines and is in line with Gazpromregiongaz and Promgaz estimates of potential energy savings in the medium and low-pressure part of the distribution system.

Table 27 provides various estimates of the rate of methane emissions from Russia's gas distribution network. There is a wide range of estimates of methane emissions of ±80%. It points out the narrow set of measurements undertaken to date which does not fully reflect the complexity, different age and state of repair of distribution networks across Russia. The average conservative rate of technological methane emissions is about 3% of the volume of distributed gas. This includes leaks and accidents but excludes commercial losses. Based on this, the overall natural gas losses from Russia's gas distribution network in 2004 can be estimated at about 5.3 bcm or 80 $MtCO_2e$.[140]

Based on this estimate of gas losses, potential gas savings in Russia's gas distribution system can be conservatively estimated at 3.5 bcm per year of natural gas or about 50$MtCO_2e$ of CH_4 emission reductions. This is a conservative estimate given that these gas savings would already be obtained if the normative emission rates were respected.

---

139. However, for this measurement exercise, details are not available on type and scale of inspected facilities.
140. This matches expert estimates of CH_4 emissions of 5 bcm for the Russian distribution sector in 1998 (IEA, 2002).

Table 27   Estimates of methane emissions along Russia's gas distribution system

Source	Emission rate, % of distributed gas	Gas losses, bcm	Methane emissions, MtCO$_2$e
Normative losses			
Gazprom normative emission rates	0.9 - 2.2%	1.5 - 3.8	23 - 56
GiproNIIgaz minimal rate of gas losses	0.6%	1.0	15
Measurements and estimates			
30% of Promgas' estimates of energy saving potential	1.5 - 3.0%	2.6 - 5.1	38 - 77
Extrapolation of Kaliningrad emission factors	5.8%	9.9	148
Kusrk measurement	0.9 - 1.1%	1.5 - 1.9	23 - 28
Tver measurement	1.5%	2.6	38
Average rate	**3.2%**	**5.3**	**80**
Range	*0.6 - 5.8%*	*1.5 - 9.9*	*15 - 148*

Sources: Gazprom, 2001a; Karasevich and Terekhov, 2004; Kuraev and Safonov, 2005; RREC, 2004; Vasiliev, 2005; IEA estimates.

GHG EMISSION REDUCTION POTENTIAL IN RUSSIA'S GAS DISTRIBUTION NETWORK

Main options to reduce GHG emissions

International experience

There is a plethora of information on projects and methods of reducing methane emissions in the gas distribution sector available through various voluntary partnerships and programmes. These include:

■ The Methane to Market Partnership international initiative.[141]

■ The Natural Gas STAR Programme of the US EPA in the North America.[142]

■ The Eurogaz-Marcogaz European working group on methane emissions.

■ The gas distribution companies participating in the Voluntary Climate Change Challenge and Registry (VCR) in Canada.

International experience indicates that given the numerous possible sources of leaks, cost-effective emission reduction programmes need to be specific to the gas distribution network of each company. A description of the measures used by the Canadian gas distribution company, Enbridge, is provided as an example in Box 8.

---

141. Since 2004, this international initiative (15 countries including Russia) focusses on advancing cost-effective, near-term methane recovery and use as a clean energy source (Methane to Market Web site).
142. This programme includes about 50 partners representing 60% of gas distribution in the United States.

Box 8   Measures to reduce fugitive methane emissions in Canada's gas distribution network

Canada's Enbridge Gas Distribution Inc. has more than 1.6 million residential, commercial and industrial customers, which consumed over 12 bcm of natural gas in 2003. In 2004, Enbridge reported the main steps it took to reduce fugitive emissions by 32% compared to 1990 levels, representing a savings of 6.1 Mm^3 of natural gas or 86 800 tCO$_2$e.

About 80% of this was achieved through the replacement of over 1 000 km of cast iron pipes with corrosion-free polyethylene pipes. A further 20% of these reductions were obtained by the "Process Venting Reduction Programme", which included the replacement of old leaking components (odorant pumps and pneumatic instruments) and the use of low-bleeding pipeline repair practices.

Source: Enbridge, 2004.

The cost of finding and fixing leaks as reported by Natural Gas STAR (2003c) ranges from USD 20 to more than USD 1 200 depending on the facility size and type of repair for volumes of gas savings up to 17 thousand m^3 per repair. Annual savings are estimated between USD 50 to USD 1 000 per repair, depending on survey costs, leak rates and number of sites.[143] These estimates are based only on the value of gas saved and do not include any market value for the related methane emission reductions. If included, these measures would be more economic and allow for the implementation of a larger number of leak reduction projects.

The Natural Gas STAR (2003c) states that the Directed Inspection and Maintenance (DI&M) programme is a "proven, cost-effective way to detect, measure, prioritise, and repair equipment leaks to reduce methane emissions". The DI&M focusses in particular on gate stations and surface facilities, which have a large number of components susceptible to leaks.

The DI&M approach is similar to that described in Chapter 3 relating to CH$_4$ emission reduction projects at compressor stations in the gas transmission system (see Table 18). A baseline survey is needed to identify and quantify leaks after which the cost-effective repairs are implemented. The leaks are prioritised by undertaking a cost-benefit analysis to estimate the value of the natural gas saved against the cost of labour, equipment downtime and the component part(s) needed to repair the leak. The final step is to develop a survey plan for future DI&M that targets components most likely to leak in the future.

The first step of the DI&M can be used to develop a well-documented baseline for Kyoto-related projects necessary for calculating achieved emission reductions. A similar approach, based on practical measurement programmes is currently used in

---

143. These estimates assumed natural gas prices in North America of USD 106/thousand m^3.

the first Project Design Document for a JI project proposal in the gas distribution system in Kursk in Russia (Kursk JI PDD, 2005).

Improved information from projects on leak rates of different equipment types could also be used to develop indicative sectoral emission factors for similar types of facilities and pipelines. In the short term, separate baseline surveys may be more suitable for JI and GIS projects given the lack of emissions data. In the longer term, JI or GIS could provide an attractive framework to implement these surveys and measurement programmes, which would contribute to developing comprehensive specific emission factors.

CH_4 emissions are not included in the existing system of emissions trading as there are no comprehensive and comparable methane emission inventories and monitoring guidelines available. For example, the EU ETS will not cover all types of GHG emissions before 2008.

Gas companies are actively working on establishing a GHG inventory, including CH_4 emissions. In Europe, individual companies are working on this under Eurogas-Marcogas. This co-operative effort may facilitate the implementation of methane emission reduction projects in the framework of flexibility mechanisms of the Kyoto Protocol.

Progress in climate-related activities in Russia's gas distribution sector

Since the first pilot project in Kaliningrad, the RREC has been actively working with Rosgazifikatsia in preparing its participation in the Kyoto flexible mechanisms. In 2004, Rosgazifikatsia created a special subsidiary company for managing fugitive emissions of its gas distribution system called "Centergazservice-opt". In 2005, the National Methane Centre (NMC) was established by Rosgazifikatsia, Centergazservice-opt, Gazpromregiongaz and RREC. The objectives of these organisations are to establish an effective accounting procedure for gas consumption and to develop a comprehensive inventory of methane emissions for all regional Russian gas distribution organisations.

Development of national guidelines for CH_4 emission reduction in gas distribution

Centergazservice-opt and the National Methane Centre plan to extend the leak measurement and repair programmes to regions with extensive gas distribution networks. This will provide a more representative base for voluntary accounting and establishing an inventory of GHG emissions than the three pilot projects implemented today.

The ultimate objective is to develop a national methodology for measuring CH_4 emissions in the gas distribution sector which would be approved by the Russian authorities. At a later stage, the national methodology for measuring emissions could be used at 79 key Russian gas distribution systems. To date, due to lack of funding and uncertainties concerning climate policy in Russia, this exercise has only been performed in the Kursk and Tver regions[144], as well as a training programme on measurement methods in Saratov for officials of distribution companies.

---

144. In 2006, a Japanese measurement methodology was tested in the regional distribution network of Tver.

Methodological work is being implemented in co-operation with Gazpromregiongaz and Russian authorities responsible for approving a national-level monitoring system of the gas distribution network. This involves Gosgorteknadzor, the Ministry of Natural Resources, and the Ministry of Industry and Energy. The extensive methane emission measurement programme will be largely based on DI&M principles. A comprehensive and accurate GHG inventory for Russia's gas distribution system would help estimate baselines for Kyoto-related projects in this sector and would also improve the quality of the national GHG inventory. Furthermore, Russia-specific emission factors could be developed and would make possible the use of a "control group" approach to define baselines for project-related activities (Kexel, 2005). This approach uses as a baseline the emission factors measured in other comparable parts of the gas distribution system outside the JI project area.[145]

Centergazservice-opt is interested in attracting investments to refurbish and modernise this sector in the framework of JI and/or GIS. Given the small-scale energy-efficiency projects in gas distribution systems, and the possible burden of transaction costs, Centergazservice-opt is concluding agreements across Russia with the largest subsidiaries of Rosgazifikatsia in an effort to streamline investments and bureaucratic procedures. Through these agreements, Centergazservice-opt's objective is to pool and manage climate-related investments and become the interface for foreign investors/buyers. This could reduce project risks and transaction costs.

Methane emission reduction options in the Kursk region: role of carbon finance

The regional gas distribution company OAO "Kurskgaz" operates over 7 170 km of gas distribution pipelines and supplies about 2 bcm of natural gas and 20 000 tonnes of liquid petroleum gas (LPG) to more than 270 000 households, about 1 800 institutions and 230 industrial companies (Kurskgaz, 2005; Kursk JI PDD, 2005). About 10% of Kurskgaz' pressure regulating stations and gas distribution mains (600 km) have passed their normal technical lifespan of operation (40 years).

The Kursk region pilot project includes 14 emission reduction options based on international experience and local practices (see Table 28). The project was due to be implemented during 2005-07, but has been unable to attract the required USD 1 million investment. Despite this, option 8 (replacing valves using teflon seals) was selected as the technological basis for the Project Design Document (PDD) for a JI proposal. As described in Table 29, the combined total natural gas savings from the 14 emission reduction options are in the order of 7.87 Mm^3 per year or 0.4% of the total gas distributed by Kurskgaz. At current domestic gas prices of about USD 40/thousand m^3, revenues would be USD 315 thousand. Some projects are more attractive than others and all projects would clearly benefit from adding the value of the related methane emission reduction credits. The economics of these projects also improves if domestic gas prices increase.

---

145. This approach has been successfully used in demand-side management (DSM) projects to calculate energy savings. Proven methodologies are available for selecting and surveying suitable control groups (PCF, 2000).

Table 28   Options to reduce methane emissions in the Kursk regional gas distribution network

Option	Number of units	Cost, USD thousand	CH$_4$ emission reduction, per year	
			million m^3	thousand tCO$_2$e
Technical options / Equipment acquisition		**194**	**1.7**	**25**
1 The use of micro-compressors to purge regulating stations with air	3 500 connecting tubes, 13 micro-compressors	17.6	0.3	3.7
2 Installation of ball valves for repair purposes (reduce length of off-service section during repair)	35 ball valves	42.2	0.3	4.4
3 Repair of gas mains under pressure without blowdown	4 repair devices	31.6	0.3	4.4
4 Automatic cathodic protection of mains	22 devices of cathodic protection	42.2	0.4	5.9
5 Supplemental diagnostic equipment to detect leaks at 40-year old mains	5 diagnostic devices	4.2	0.1	1.8
6 Acquisition of new excavators improving capacities to repair gas facilities	5 excavators	56.2	0.3	4.4
Replacement		**717**	**4.9**	**72**
7 Automatic control of gas parameters and leaks at pressure regulating stations	3 500 system of automatic control	105.4	0.6	8.1
8 The use of teflon seals at gate valves and flanges	6 600 gate valves 20 000 flanges	232.6 8.1 240.7	3.0	44.1
9 Replacement of pressure regulating stations over 40-years old	1 pressure regulating station	26.7	0.4	5.1
10 Installation of plastic mains and service pipelines (replacement / construction)	8 welding facilities	274.1	0.5	7.4
11 Replacement of trilinear manometer plugs by the T-shaped connectors	7 000 T-shaped connectors	70.3	0.5	7.4
Organisational		**123**	**1.3**	**19**
12 Use of 10 additional repair crews for simultaneous maintenance needs	Seconding 10 crews & related equipment	105.4	1.0	14
13 Optimisation of repair works using statistical models	Development of statistical models	7.0	0.2	2
14 Optimisation of maintenance practices using projections of statistical models	System of automatic management of repairs	10.5	0.2	3
Total		**1 034**	**7.87**	**116**

Source: Kursk methane emission reduction options, 2005.

The pilot project in Kursk encompasses both technical and organisational aspects described in Table 28. It demonstrates the close link between financial and technical barriers of regional and municipal gas distribution organisations. Options 1 to 11 include the replacement of existing facilities as well as the improvement of diagnostic equipment, maintenance and repair practices. Investing in new equipment, such as diagnostic devices, excavators and cathodic protection devices (options 4-6) is also essential in order to cope with the growing number of repairs required on ageing equipment. Options 12 to 14 reflect the lack of human resources to undertake repairs and conduct timely and effective preventive diagnostic work.

If JI is implemented, the demonstration of *additionality* of these CH_4 emission reduction options would need further assessment. This demonstration should take into account the existing economic and financial incentives for gas savings by Russian gas distribution organisations. In this regard, the approved CDM methodology for a Moldovan project to reduce leaks from natural gas pipelines, compressors or gate stations (CDM EB, 2005b) can be useful. This methodology provides different possible outcomes of additionality test depending on whether or not companies are penalised for lost gas or rewarded when these losses are reduced (see Box 9 on the "cost-plus" tariff formula):

■ If financial incentives exist, investment analysis should categorise all economically attractive options as non-additional.

■ If financial incentives are limited, the barrier approach may be used to demonstrate impediments to implementation of otherwise economically-attractive options in the baseline scenario.

Economics of business-as-usual and Kyoto-related investment in gas savings

The implementation of the complete set of proposed project options in Kursk could reduce natural gas losses by 7.87 Mm^3 per year equal to 116 $ktCO_2e$ by the end of the project implementation from 2005 to 2007 (see Table 28). In order to estimate the economic attractiveness of these CH_4 emission reduction options, the following assumptions were used:

■ A discount rate of 25% reflecting the high risk of projects in Russia, in particular in the regional gas distribution sector.

■ The regulated price of natural gas in the Kursk region of USD 40/thousand m^3.

■ A 7-year crediting period in which the reduction of methane emissions can be accumulated, based on CDM EB (2005b). [146]

The bulk of options have no net cost to reducing methane emissions over a 7 year crediting period due to the value of saved gas. Given the quick return on investment of these options, discount rates do not especially affect project economics. The accumulated reductions of CH_4 emission of the package of 14 proposed options over the 7 year period is over 55 Mm^3 or 0.8 $MtCO_2e$. However, each option by itself provides a relatively small part of this total – about 50 000 tCO_2e per option (see Figure 22).

---

146. This period can be prolonged for another 7 years with necessary baseline adjustments.

Figure 22   Estimated costs of methane emission reduction options in the Kursk region

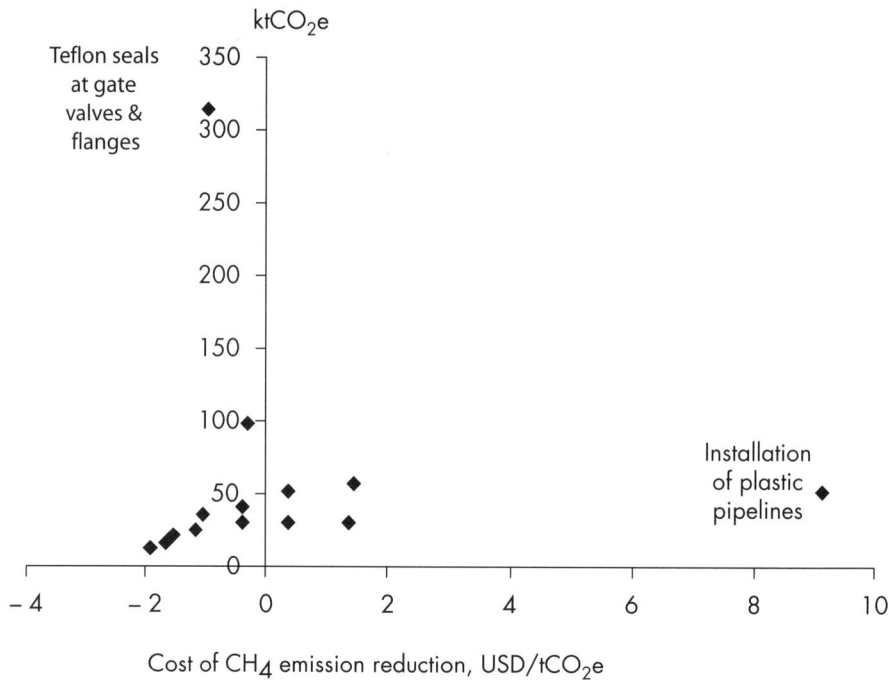

Source: Kursk methane emission reduction options, 2005; IEA estimates.

Two gas-saving options clearly stand out:

■ Option 8 (to replace gate valves and flanges by teflon seals) will generate large reductions of CH_4 emissions at no net cost, given the value of saved gas. This option was also used in Kaliningrad (see Box 7) and was chosen for JI.[147]

■ Option 10 (to replace old mains by plastic pipelines) is the most expensive option in the Kursk pilot project. For this reason, it ranks as a "second or third order option", requiring more financial investment and a longer pay-back time.[148] However, its emission reduction cost is still attractive, just over USD 9/tCO_2e.

Given the current rules used in Russia to calculate regulated tariffs (see Box 9), Kurskgaz will have an incentive to invest in projects with pay-back times shorter than 3 years. If methane emission reductions are monetised at 7 USD/tCO_2e, all projects (except the project replacing mains) have a pay-back time of less than 3 years (between 0.4 and 1.8 years). Without carbon revenues only 6 of the 14 options have a pay-back time of less than 3 years, generating only half of the total expected leak reductions (see Figure 23).

---

147. The Kursk JI PDD (2005) reports a much larger emission reduction from this measure of about 300 ktCO_2e by the time all replacement activities are implemented. However, it indicates that both project and baseline scenario emissions will only be calculated once the project starts.
148. We use here the criterion of pay-back time in order to facilitate comparison with the 3-year period allowed for investment return used in the regulated gas tariff formula in Russia (see Box 9).

Figure 23   Estimated pay-back time for methane emission reduction options in the Kursk region

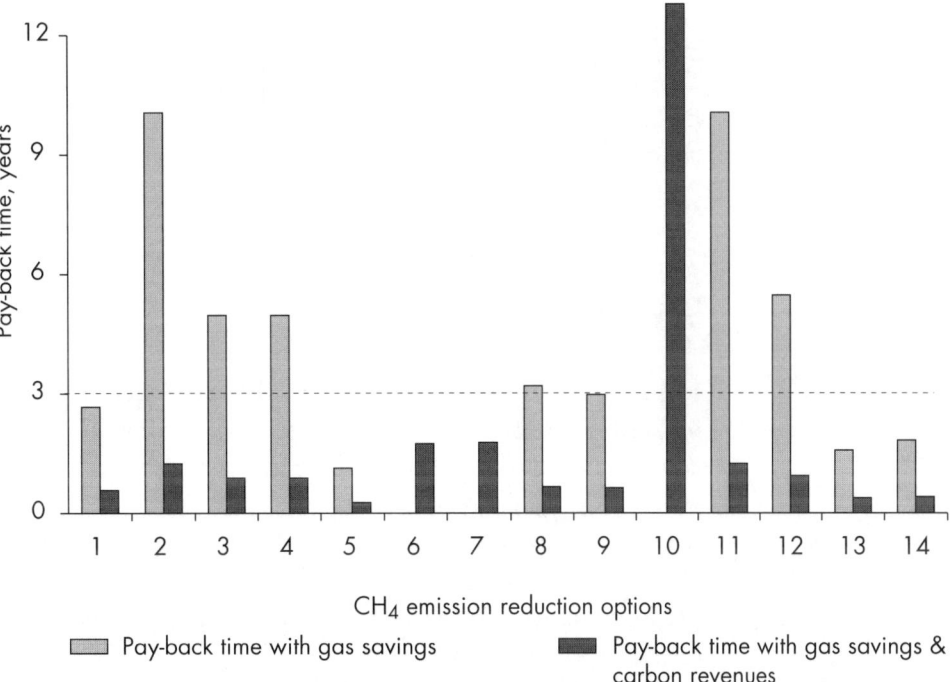

Source: Kursk methane emission reduction options, 2005; IEA estimates.

Carbon finance could also facilitate access to capital by Russian gas distribution companies which may have difficulty using normal types of financing such as: issuing new equity (hindered by their state-owned nature), using bonds (hindered by the small-scale nature of projects); and taking out bank loans (hindered by being unable to use state equity as collateral). Carbon finance could be a source of finance with potentially lower costs, accessible to foreign investors/buyers or be considered as a future guaranteed source of revenue for the project. In most cases, however, carbon finance does not cover the entire amount of the investment, but only the specific cost of emission reductions or their negotiated purchase price. This may be of less importance for the majority of gas distribution companies increasingly under the control of Gazprom.

Gas distribution companies in Russia may also find it difficult to make economic and financial evaluation of projects due to lack of experience. Kyoto-related projects could, therefore, support gas-saving projects by providing stimulus through funding, as well as capacity building, and contribute to more efficient operations.

The attractiveness of Kyoto-related projects could be hampered by transaction costs related to the JI project cycle. For projects aiming to reduce methane leaks, the measurement of hundreds of kilometers of distribution network, can make it difficult and costly to verify emission reductions. Monitoring and transaction costs of these individual small-scale projects may be reduced if the developer could bundle projects together.

According to the International Emissions Trading Association (IETA, 2006), existing estimates of CDM transaction costs range widely between USD 40 000 for simple projects to USD 400 000 for complex projects. For the purpose of our analysis, we have assessed the impact of transaction costs at the lower-end of this range, integrating costs in the order of USD 40 000 and USD 80 000 to the various options described in the Kursk project (see Figure 24).[149] As expected, these transaction costs significantly increase the costs of CH_4 emission reductions. Transaction costs of this order have a relatively stronger impact on the less expensive projects, more than doubling their CH_4 emission reduction costs. In contrast, the more expensive projects (replacement of mains, for example), can more easily support this level of transaction costs.

It is important to develop standardised baseline criteria and streamlined procedures for small-scale projects in the gas distribution sector. JI procedures may be too complex and costly for this type of project, especially under the JI Track 2 unless the principle of project bundling can be used.[150] The GIS may be a "friendlier" framework, provided that small-scale options can be bundled. A GIS "programme approach" whereby a group of similar small-scale projects can be developed together (see Box 2) or sectoral emission factors can be used, may be necessary to implement such small projects.

Midterm "common practice" in Russia's gas distribution system

Programmes already put in place by Gazprom could be used as an indication of "common practice", to facilitate the development of baseline scenarios for Kyoto-related projects in Russia's gas distribution sector. This can provide the basis on which to show the additional nature of Kyoto-related projects over what is considered to be "common practice" in a given sector (see Box 4).

The notion of "common practice" may also be useful for the authority responsible for approving JI projects and/or GIS. It can reflect historic practices (over the past 5 years, for example) and/or comprehensive projections of technical developments of gas distribution networks in the medium-term. If easily available, information related to the "common practice" can facilitate the selection of projects and reduce costs for project developers and authorities in charge of JI/GIS project approval.

This approach, however, must be used with caution, given the "asymmetry of information" between the owner and the user of information (the responsible authority). In theory, a company's behavior and its projections could be influenced by its interest in obtaining more carbon revenues. In other words, it could underestimate or slow down efforts to reduce GHG emissions for the purpose of maximising possible carbon revenues in the future. Clearly, this situation would be unsatisfactory, both

---

149. These types of projects may be more complex and have higher transaction costs than our assumption reflects.
150. At the time of publication, this was being discussed by the Joint Implementation Supervisory Committee.

Figure 24   Sensitivity of methane emission reduction options to transaction costs

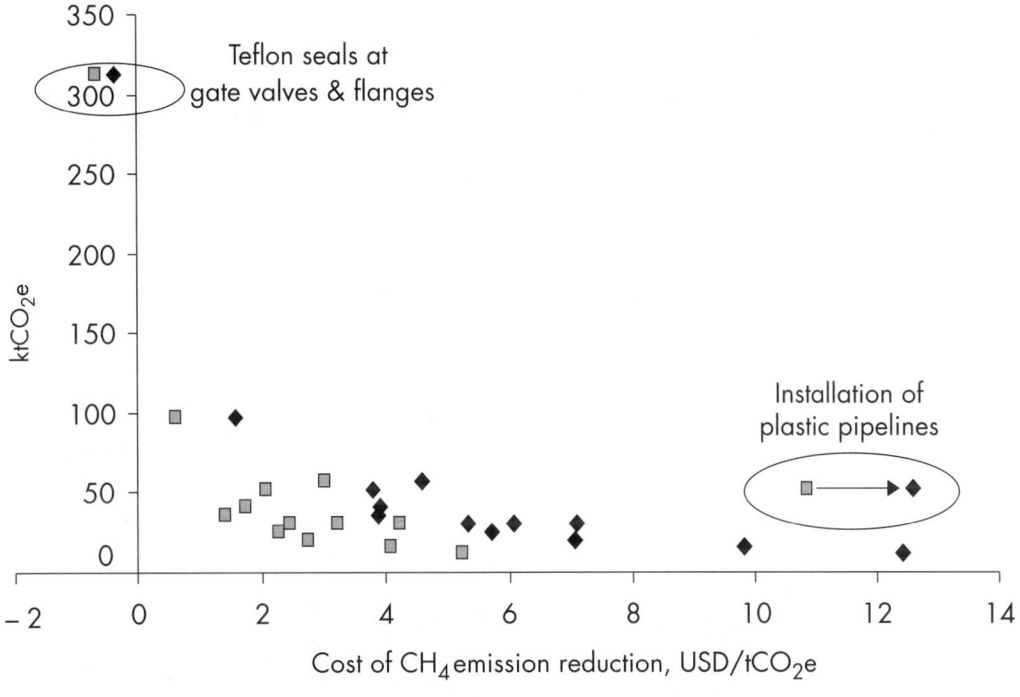

Source: Kursk methane emission reduction options, 2005; IEA estimates.

from an economic and environmental point of view. On the other hand, taking into account that gas distribution organisations in Russia have to justify the investment component in the regulated gas tariff, the underestimation of investment needs may not be in the company's interest. Establishing clear objectives and guidelines for companies participating in Kyoto Protocol flexibility mechanisms in Russia may make companies more open and enhance incentives to make such projects possible.

Gazprom's 2004 Programme of "Reconstruction and Technical Modernisation of the Gas Distribution Network" provides a preliminary basis on which to assess what could be considered "common practice" in Russia. This Programme establishes common technical practices for its gas distribution subsidiaries (Gazprom, 2004c). As Gazprom's control over the gas distribution sector increases, its Programme may become representative of Russia's gas distribution sector as a whole from which elements of common practice can be drawn.

The objectives set out in Gazprom's Programme confirm that extensive investigation and monitoring are still necessary to determine the current technical state of Russia's gas distribution system. By 2005, Gazprom had planned to develop a sectoral system to enhance energy savings, and reported the establishment of an inventory of its gas distribution facilities. However no further details are provided apart from information on the implementation of two projects in Astrakhan and Tver, related to the efficient gasification of regions.[151]

In 2001, Gazprom developed a list of priority measures to control and mitigate GHG emissions in the gas distribution sector for 2001-12. Total potential reduction of CH_4 emissions is projected to be 1 bcm/y or about 15 $MtCO_2e$ per year by 2012 (Energy Security of Russia, 2005). The most significant measures to reduce gas losses by 0.8 bcm/y (12 $MtCO_2e$) are:

■ Comprehensive use of modern metering equipment for industrial, residential, commercial consumers and boilers. This measure could reduce the "imbalance" and be an incentive for more efficient gas consumption.

■ The monitoring, detection and repair of leaks on mains, valves, metering and regulating facilities of gate stations and pressure regulating stations as well as on equipment installed for residential consumers.

Two other options, leading to less significant gas savings, include:

■ The use of inert gases in the blowdown of mains and repair technologies without having to disconnect from gas mains could lead to the saving of 0.16 bcm of natural gas per year.

■ The monitoring of regulating facilities (maintaining pressure) at pressure regulating stations and the replacement of older regulating facilities could save 0.12 bcm/y of gas.

Gazprom's Energy-Saving Programme for 2001-10 is the primary source where Gazprom describes its potential options and estimates of costs (Gazprom, 2001a). It considered all but one of the above opportunities as low-cost options at less than USD 17/thousand m³ of saved gas ("first or second order" options). The installation of no or low-bleed metering equipment along gas distribution pipelines was considered as a high-cost measure. It estimated the cost of this measure at more than USD 160/thousand m³ of natural gas saved and delayed its implementation until after 2010.[152]

The increase in domestic gas prices has enhanced the project economics of many potential investments in this sector. Whereas in 2001 when domestic natural gas prices were at USD 15/thousand m³, the installation of metering equipment along gas distribution pipelines was not economically attractive. At current domestic prices, however, this project could be of interest for Kyoto-related investments, as the cost of emission reduction is below USD 15/tCO_2e.[153]

---

151. One energy-saving project implemented by OAO Promgaz (the planning and design centre of Gazprom) in the Astrakhan region includes its gasification, the energy audit of end-users and the development of investment proposals to increase energy efficiency. The second project is the establishment of a high-energy efficiency demonstration zone in the Tver region and is being implemented jointly by OAO Promgaz and Ruhrgas.
152. Gazprom estimated total project costs at USD 18.5 million and annual gas savings of 110 Mm3.
153. Furthermore, integration of potential carbon revenues could render projects more attractive at current prices, and even more so, at future domestic gas price levels. The possibility to access capital at lower costs, further decreases emission reduction costs to only USD 1.4/tCO_2e.

However, the use of the "common practice" approach to assess the additionality of Kyoto-related projects in the gas distribution sector would need to take into consideration the existing barriers for gas distribution companies in Russia that could prevent the implementation of these practices. Beyond the barriers posed by the domestic tariff level, two main factors limit incentives for a gas distribution company to invest in gas savings: the regulated tariff structure which does not include incentives for longer-term investments in gas savings and the ownership structure (lack of competition and the lack of separation between management and regulatory functions).

Financial incentives to gas saving: tariff structure

As well as having to deal with regulated tariffs which do not fully reflect costs, gas distribution companies in Russia may also have little to no incentive to make capital investments in gas savings, namely if this investment aims to reduce operating costs (*i.e.* reduce leaks). This is because the company is not sure to recoup its investments through the regulated tariff structure. In Russia, regulated tariffs use a "cost plus formula" (see Box 9), which gives little incentive to reduce costs. For a gas distribution company costs are related to total volumes of transported gas including losses. Therefore it has little incentive to reduce costs given its profits are directly tied to throughput volumes.

Box 9   "Cost-plus formula" for tariffs in gas distribution in Russia

According to the Methodological Principles of Tariff Regulation for gas transport and distribution services (Federal Energy Commission of RF, 2003), the average tariff is calculated as:

$$T = \frac{(Costs_{transport} - R_{operational} - R_{other} + Taxes + Net\ profit) \times (V_{final\ cons.} + V_{transit})}{V_{\Sigma}^{2}}, \text{ where}$$

$Costs_{transport}$ – is the sum of transportation costs included in the cost calculation,

$R_{operational,\ other}$ – net revenues of transportation and other activities,

V_{Σ} - total throughput of the gas distribution system, calculated as:

$$V_{\Sigma} = V_{transit} + V_{final\ cons.}, \text{ with}$$

$V_{transit}$ – volume of gas transit in the throughput of gas distribution system. This volume does not include the gas consumed for technological needs of the gas distribution company and the gas used to compensate losses due to accidents.

$V_{final\ cons.}$ – volume of gas distributed to the final consumers. This component of the tariff renders the rewards for gas savings even more complex, if one takes into account the lack of metering equipment, especially at the end-use level.

However, the capital cost component of regulated tariffs in Russia remains stable for a three year period. This three-year limit provides some incentive for utilities to invest in efficiency measures to reduce costs (due to gas savings, for example), but favors projects with pay-back times of less than three years. This is often too short for major capital-intensive efficiency investments.

In future, general tariff-regulating principles should evolve in the direction of incentive-based regulation, as used in several countries of Central Europe, and in Sweden, for example. IEA (2004a) discusses in more detail the issue of tariff regulation in the case of district heating, which could be applied to the gas distribution sector. Predictable and robust incentive-based regulation, such as price caps or benchmarking[154], can ensure that the operator has sufficient motivation to improve efficiency while keeping the benefits for a long-enough period to make a return on its investments.

Financial incentives to gas saving: ownership structure

Motivations for investment by municipal (state) owners of gas distribution networks may differ from those that drive private owners. A municipality may have an objective to keep tariffs low for social or political reasons (re-election, for example) and not to give priority to investments in energy savings, which could be initially reflected in higher tariffs.

By analogy with district heating systems, changes in the ownership structure, essentially the separation of management decisions from political considerations and the introduction of competition between gas distribution companies, could provide more incentives to increase efficiency (IEA, 2004a). This could include the transfer of certain management/operation tasks to private companies using different forms of ownership from the point of view of the responsibility for investment and risk sharing – from short-term service contracts to privatisation of assets. For example, a concession may include the obligation to undertake technical improvements to the system and improve the quality of service thus providing a direct incentive to the operator to improve efficiency and reduce costs.[155]

The consolidation of distribution companies by Gazprom is currently a major restructuring trend in Russia's gas distribution sector. As argued by Gazprom, this can enhance the financial capacities of distribution companies. At the same time, it is questionable whether or not efficiency investments will be given priority in comparison with investments in new regional gasification programmes.

Overcoming barriers using Kyoto Protocol flexibility mechanisms

This brief analysis of methane emission reduction options related to Gazprom's Programme, as well as those proposed by Kurskgaz, confirms that the Russian gas distribution sector has many "first and second order" economic projects. These

---

154. Benchmarking or "competition by comparison" induces utilities to compete with one another for cost savings, even when they are not operating on the same local market.
155. For example, timely maintenance, response to complaints, etc.

options could be implemented during maintenance and repair programmes and encompass part of common practice in this sector. However, the limited financial and technical capacities of gas distribution organisations are a barrier to implementing such practices.

Kyoto Protocol flexibility mechanisms could be extremely useful and timely in helping overcome these barriers, and attract much needed investment to this sector. The use of carbon finance could improve awareness of current levels of gas losses and stimulate gas savings. Kyoto-related projects, if implemented jointly with foreign partners, could also bring necessary knowledge and best practices to improve current maintenance and repair performances of Russia's gas distribution companies. However, the Kyoto flexibility mechanisms cannot be viewed as a unique solution to the existing problems and barriers that limit the efficient functioning of the gas distribution sector. Sector reforms, including price and regulatory reforms, are essential in removing existing financial and technical barriers to more efficient gas consumption throughout the gas distribution sector and by all Russian end-users and consumers.

V. REDUCING GAS FLARING: OPPORTUNITIES AND CHALLENGES

GAS FLARING AROUND THE WORLD

Associated gas is a byproduct in the production of oil as it is brought from high pressure in the reservoir to low pressure at the surface. Dissolved gas comes out of this solution, similar to opening a bottle of champagne. Associated gas is of different composition and quality varying widely in its content. It usually has a lower methane content than non-associated gas, but is still a valuable fossil fuel, very similar to natural gas. The content of associated gas in the oil is usually expressed as a Gas-to-Oil Ratio (GOR), a volumetric ratio of gas to oil at surface conditions. GORs vary widely in different reservoirs around the world from about 10 to several thousand cubic meters of gas per cubic meter of oil. Hydrocarbon deposits with very high GORs are usually called gas condensate fields rather than oil fields, and are exploited for their gas.

Why is such a valuable fossil fuel flared? In some cases, flaring is for safety purposes, such as during emergency shutdowns or disruptions in processing systems to release dangerous pressure build-up. In some cases "venting" can occur, where the gas is not flared but released directly into the atmosphere. However, the practice of venting is often restricted by regulations due to safety concerns. Operators prefer flaring associated gas for the same reasons of safety. In other cases, as we will see below, associated gas is flared because alternative uses are uneconomic often due to the long distances between production and consuming centres, or to geophysical difficulties related to re-injecting the associated gas into the field without negatively affecting oil recovery. Associated gas is also flared due to market failures, often because of the market structure that creates barriers to investment or access to the necessary infrastructure. Whatever the reason, routine flaring to dispose of associated gas is a waste of energy resources and contributes to a considerable increase in GHG emissions released into the atmosphere.

The Global Gas Flaring Reduction Partnership (GGFR) lead by the World Bank (see Box 10) conservatively estimates 150 bcm of associated gas were flared worldwide in 2004 (World Bank, 2006).[156] This is roughly equivalent to 5.5% of the world's total gas consumption, or 30% of the European Union gas consumption in 2004.

---

156. This number was derived from the data officially reported by GGFR partners, while for non-partner countries GGFR primarily used data published by Cedigaz.

Box 10   Global Gas Flaring Reduction Public-Private Partnership

The Global Gas Flaring Reduction Public-Private Partnership (GGFR) was launched at the World Summit on Sustainable Development (WSSD) held in Johannesburg in 2002. It was previously known as the Global Initiative on Gas Flaring Reduction launched by the government of Norway and the World Bank Group. The GGFR aims to support national governments and the petroleum industry in their efforts to reduce flaring and venting of associated gas.

The current members of the GGFR include: governments from oil-producing countries (Algeria, Angola, Cameroon, Chad, Ecuador, Equatorial Guinea, Indonesia, Kazakhstan, Nigeria and Qatar), international oil companies (BP, Chevron, ENI, ExxonMobil, Marathon Oil, NorskHydro, Royal Dutch Shell, Statoil, and TOTAL), the World Bank Group, the OPEC Secretariat, and donor countries Canada, Norway, the United Kingdom, and the United States. The GGFR partnership represents close to 70% of global flaring. In 2004, the regional government of the Khanti-Mansiysk, the biggest oil-producing region in Russia, became a member of the GGFR.

The GGFR developed the "Global Gas Flaring and Venting Reduction Voluntary Standard" which was unveiled at its second international conference in Algeria in May 2004 (GGFR, 2004b). The implementation of this Standard aims to cut venting and flaring significantly within 5 to 10 years in the GGFR partnership. The Standard provides a framework for companies, governments, and other key stakeholders to encourage joint rather than individual actions to identify and evaluate economically feasible alternatives to gas venting and flaring. It also seeks to create a supportive gas infrastructure and gas market. Other activities include data gathering, stakeholder consultations, and identification and dissemination of best practices.

The GGFR work programme initially focussed on: i) commercialising associated gas, including domestic market development and access to international markets; ii) developing legal and fiscal regulations for associated gas; iii) implementing the GGFR Voluntary Standard for Global Gas Flaring and Venting Reduction; and iv) capacity building related to carbon credits for flaring and venting reduction projects. The focus has evolved to be more country-specific, with an emphasis on demonstration projects to lead the flare reduction process.

The GGFR releases reports highlighting international experience and GGFR work such as opportunities to use the JI and CDM for flaring reduction projects (GGFR, 2005b), opportunities for small-scale use of gas (GGFR, 2004c), and the regulation of associated gas flaring and venting, including regulatory profiles of countries (GGFR, 2004a).

Source: GGFR, 2005a.

In estimating this impact in terms of GHG emissions one must take into account the efficiency of the flare (see Annex 1). Assuming a global flaring efficiency of 98%, the World Bank's global flaring estimate corresponds to 435 MtCO$_2$e. This represents about 9% of the estimated allowances needed by Kyoto Protocol Parties to meet their emission targets over 2008-12.

During the 1990s, there was a move toward more stringent regulation of flaring which has been combined, in some cases with ambitious phase-out targets. Countries have adopted a variety of policy instruments to address flaring within their borders (*e.g.* various regulatory approaches and practices, taxes and emission fees, and negotiated or voluntary agreements). The effectiveness and cost-efficiency of these policies have varied significantly between countries. Moreover, as pointed out by the GGFR (2003), measures to regulate flaring cannot be isolated from broader policies pursued by governments to promote efficient gas use, especially through reforms of the gas sector.

Economics of gas flaring

Ideally, oil companies would like to monetise their associated gas production. This requires either having a local market near the producing field, or having to transport it long distances to a market. Often, in remote areas – as illustrated later in this chapter in our "typical" example for West Siberia – there is not a large enough local market to consume a significant share of the associated gas produced. Also the total amount of gas is often not deemed sufficient to justify the capital investment (pipeline, LNG plants and tankers, etc.) needed to transport the gas.

Figure 25 presents the various options available for using associated gas as a function of the volume produced and the distance to markets at current gas prices and state of technology. For volumes less than 10 Mm^3/day and distances to market greater than 2 000 km, all options are currently uneconomic. In this case, environmental policies or fees are essential to prevent gas flaring. At distances less than 2 000 km, a range of options can be potentially economic. Associated gas can be used to generate electricity or be transported by pipeline to consumption points. At volumes above 10 Mm^3/day and longer distances to market, liquefied natural gas (LNG) or gas-to-liquids (GTL) projects may make economic sense. Some Russian oil companies are active at the R&D level in GTL projects, suggesting that this might be an attractive possibility in the future.

If a downstream gas market is not economically accessible, the next best option is to re-inject the associated natural gas back into the reservoir. Depending on the characteristics of the reservoir, this may be attractive because it can help increase total oil recovery. Associated gas can be re-injected in the gas cap, if there is one, for pressure support, or in certain circumstances it can be re-injected directly into the oil zone to improve drainage.[157] This is usually done in the later stages of production as an enhanced oil-recovery technique. However, in other cases, it can decrease production and recovery as the gas "breaks through" and simply gets recycled. Detailed geological studies are needed to determine whether or not oil recovery will be enhanced by this

---

157. Many West Siberian reservoirs do not have a gas cap.

process. Economic analysis is also needed given the costs involved to compress the gas to match the high pressures of the reservoir. Consequently, without enhanced oil production reaching a certain threshold rate, gas re-injection cannot be justified based on economics alone.

When the conditions are right for re-injection:

■ Re-injection will usually improve recovery, but not immediate production. In other words, the additional production is delayed. With high discount rates, this gives only minimal economic value to the additional recovery.

■ Re-injected gas is in principle not lost as it can be recovered at the end of the field life. This can be advantageous if the operator expects the value of gas to increase significantly in the future.

Currently, associated gas is never re-injected in non-reservoir layers, such as saline aquifers. This is a costly process with few economic benefits. However, if CO_2 emission reduction credits could be earned, the economics could be improved. The revenues from CO_2 emission reductions could offset the re-injection costs, making it economically neutral to simply "store" the gas in a suitable geological formation

Figure 25   The economics of alternatives for associated gas use

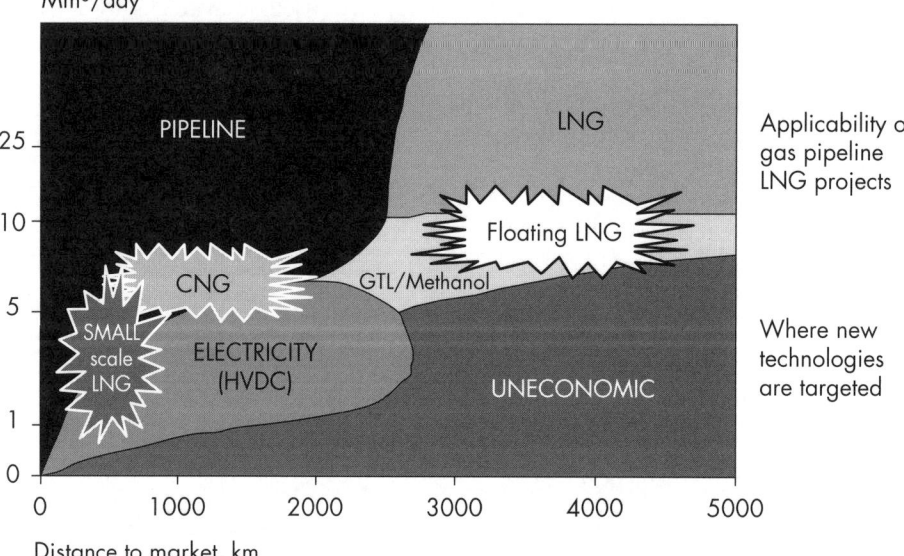

LNG: liquefied natural gas
GTL: gas-to-liquids
CNG: compressed natural gas
HVDC: high voltage direct current

Source: Reproduced with permission from SINTEF.

for future use.[158] This may be advantageous if the gas injection would increase oil production and recovery, and/or if a suitable aquifer layer exists with larger injectivity and/or lower pressure, thereby reducing re-injection costs.

The situation in Russia

There are several factors explaining the prevalence of flaring in many Russian oil fields. Economics are a key consideration. The remoteness of many producing locations (*e.g.* West Siberia) combined with limited local markets not large enough to consume the produced associated gas, often do not justify the large capital expenses to construct transport infrastructure. The structure of Russia's natural gas sector adds a key regulatory barrier to the list of purely economic ones against the monetisation of associated gas production. The lack of reliable and transparent access to Gazprom's monopoly gas pipeline infrastructure has been a constant complaint by independent gas producers in Russia over the past years.

The IEA considers that large volumes of gas produced by oil companies are still being flared because Gazprom declines to buy it, or because the terms of access to processing plants and the transmission network are uneconomic (IEA 2004b). This is consistent with the assessments made by the GGFR in their Russia-related work (GGFR, 2003).

Another important factor is that until recently Russian oil companies have been more focussed on increasing their oil production and oil export revenues and not on monetising the associated gas produced. Since 1999, Russian oil companies have managed to increase their oil production by over 1 million bbl/d. However, as oil export outlet capacity reaches its limits, and the Russian government continues to discuss various export pipeline routes, Russian oil companies are starting to focus on gas production and better utilisation of associated gas as a way to enhance their revenue stream and profit. The absence of a clear government policy to stimulate independent gas production has not provided any comfort to oil companies to lobby for what has been to date a secondary interest.

In this respect, the increase in pollution fees in 2005 levied on methane emissions above an allowed level (see Chapter 2) should increase the lobby by Russian oil companies for more reliable and transparent third party access to the Gazprom network. Clearly, the Russian oil industry will not want to be penalised for a problem beyond its control. It would seem that the more forward-looking Russian oil companies, whose size and power make them a strong lobby group, are preparing the groundwork for making such a case. As Russian domestic gas prices increase, enhancing gas production will become an increasingly attractive investment proposition for oil companies with fields close to gas pipelines and gas processing infrastructure.

---

158. Currently, there are only 4 projects in the world where CO_2 (and not the associated gas) is re-injected into saline aquifers for storage as a climate change mitigation measure. This is done in Sleipner and Snøhvit in Norway, In-Salah in Algeria and Gorgon in Australia.

ESTIMATES OF GAS FLARING IN RUSSIA

Official estimates of gas flaring

Official Russian government statistics on associated gas flared by oil companies reports a range over the 1990s from a high of 10 bcm in 1991 to a low of 4.3 bcm in 1996, a low point in Russian oil production. Since then, the increase in Russian oil production over 1999 to 2005 brought with it a significant increase in associated gas flaring. In 2005, flared gas volumes were officially reported as 14.98 bcm (see Table 29). This represents about 27% of associated gas production in 2005 in comparison with a flaring rate of 20% in 1999 (IEA, 2002).

Although the volume of flared associated gas has increased in Russia given the growing volumes of oil produced and, in some cases, due to the maturity of oil fields, many Russian oil companies are improving their gas utilisation rates. Small and medium-sized Russian oil companies, producing less than 6% of Russia's oil production, have lower GORs yet flare a higher share of produced associated gas - 45% versus the majors' 26%.

Major Russian oil companies are making use of part of their associated gas as input for their gas processing facilities and to fuel co-generation plants to provide their own and local electricity needs. Increasingly, companies are showing a keen interest in stepping up natural gas production in the coming decade as reflected in their annual reports and public statements.

Russian oil company initiatives to enhance utilisation of associated gas

Surgutneftegas

Surgutneftegas produced 28% of total Russian volumes of associated gas or 15.4 bcm in 2005 (see Table 29). It processed 13 bcm of this into dry gas, liquid hydrocarbons for domestic and municipal use as well as generating electricity for own-use and local needs. Thus, although the largest producer of associated natural gas, Surgutneftegas reports to have the lowest gas flaring ratio, at only 7% of annual production.

More than half of Surgutneftegas' environmental investments in 2004 were used to construct new gas turbine power generators. This investment is considered to be the best option for using associated gas at new oil fields situated far from natural gas pipeline infrastructure. In 2001-04, Surgutneftegas constructed eight gas turbine power plants, namely in the Khanti-Mansiysk region.

Surgutneftegas plans to invest in 6 gas turbine power plants in 2006 and to refurbish two existing plants. This will increase its gas utilisation rate above 95%. During 2007-12, new power plants are to be constructed as well as a pipeline from the Fedorovskoe field, thereby using an additional 2.1 bcm of associated gas per year. Given the substantial reduction of GHG emissions due to these investments, Surgutneftegas is considering the possibility of accessing carbon revenues via the JI mechanism. In 2005, an associated gas-to-power Project Idea Note (PIN) was prepared by Surgutneftegas in the framework of the Khanti-Mansiysk Autonomous

Table 29   Produced and flared associated gas, by oil company, in 2005

	Produced associated gas, bcm	Flared associated gas	
		bcm	% of production
Lukoil	6.15	1.38	23%
Rosneft	8.49	3.18	37%
Yukos	2.61	0.64	25%
Sibneft	5.66	3.68	65%
Surgutneftegas	15.42	1.06	7%
TNK-BP	10.70	2.40	22%
Tatneft	0.77	0.03	4%
Bashneft	0.43	0.01	22%
Slavneft	1.53	0.54	35%
RussNeft	1.56	0.50	32%
Total oil companies	*53.31*	*13.50*	*25%*
PSA operators	0.47	0.01	21%
Others*	2.49	1.38	55%
Total Russia	**56.27**	**14.98**	**27%**

* including Gazprom and Novatek

Sources: Energy Sector of Russia, 2006.

Region's partnership in the GGFR. This PIN has now been submitted to the World Bank Carbon Finance Unit. If successful, and depending on the Russian JI rules yet to be adopted, the project could become eligible for carbon finance.[159]

Yukos

Once Russia's largest oil producer, Yukos' key oil producing assets are run by the state oil company Rosneft since the end of 2004. Accounting for over 60% of Yukos production, already before events in 2004, Yuganskneftgas fields produced over 50% of Yukos' associated gas (2.9 bcm) in 2003 and accounted for about 65% of the company's flared gas (2.4 bcm). Analysis of the Yuganskneftegas fields (see Table 30) shows the GOR increased by 16% between 1999 and 2003. This shows a trend of increasing GOR as fields mature.

Yukos had a stated goal to increase its associated gas utilisation rate to 85% over the period 2003-08, from its level of nearly 60% in 2004. The problems faced by

Table 30   Production at Yuganskneftgas fields, 1999 vs 2003

	Oil production, Mt	Associated gas production, bcm	Gas-to-Oil Ratio (GOR)	Gas flared as a % of total production
1999	26.2	1.3	43	26%
2003	49.7	2.9	50	41%

Source: Energy Sector of Russia, 2005; Yukos, 2003.

---

159. Discussion with GGFR officials in 2006.

Box 11   Sibur associated gas processing activities

OAO "Sibur" *(Sibirsko-Uralskaya Neftegazohimicheskaya companiya)* – "Petrochemical company of the Urals and Siberia" – is the largest producer of chemicals from oil and gas in Russia. OAO "Sibur" was created in 1995, and by 2001 had about 60 gas processing and petrochemical plants. In 2005, SIBUR Holding was created (25%+1 share of Gazprom, 75%-1 share of Gazprominvestbank).

Associated gas produced by Russian oil majors is a key source of gas input for Sibur. In 2004, 65% of the almost 11 bcm of associated gas processed by Sibur was undertaken by its SiburTyumenGaz subsidiary. In 2005, the volume of associated gas processed was 13 bcm, a 20% increase over 2003 levels. This corresponded to less than 65% of the rated design capacity of Sibur's gas processing plants. This lower than rated throughput reflects the need for upgrading and refurbishment to push its capacity utilisation higher and increase the use of associated gas. In the Sibur region, about 6.5 bcm of associated gas was still being flared in 2004, according to official reports. In 2004, Sibur and Gazprom estimated that USD 300 million was needed to increase the volume of associated gas processing by 45%.

Source: Plotnikov, 2005; SIBUR Web site; Energy Sector of Russia, 2005.

Yukos (at its remaining fields) and now by Rosneft at the Yuganskneftegaz fields are related mostly to the complete dependence of oil producers on the capacity of the gas processing arm of Gazprom, Sibur (see Box 11), and the lack of incentives for Sibur to enhance its distribution (gas-collection) network or to put in place more gas processing plants. The fact that regional or local markets are just not large enough to consume the volumes of associated gas produced, adds yet another economic disincentive to effective gas utilisation. Table 31 reflects these challenges, showing the volume of gas flared increasing beyond or outside the zones where Sibur gas processing plants (and collector lines) are located. This is mitigated to some extent by Yukos' ability to process gas at its own plants or for its own use.

Table 31   Yukos associated gas flaring and utilisation and its dependence on Sibur

		1998	2001	2004
Oil production, Mt				
	Total	44.5	58.1	86.0
	Within Sibur zone	25.8	31.5	46.1
	Outside Sibur zone	18.7	26.6	39.9
Gas flared, bcm				
	Total	1.0	1.6	2.3
	Within Sibur zone	0.5	0.4	0.3
	Outside Sibur zone	0.5	1.2	2.0
Gas Utilisation, bcm				
	Total	1.4	1.7	3.4
	Own processing plants	0.3	0.3	1.4
	Electricity (own needs)	0	0	0.2
	Sibur processing plants	0.7	0.9	1.1
	Other	0.5	0.5	0.7

Source: Payusov, 2005.

Figure 26   Share of gas flared at Tomskneft due to Gazprom network limitations

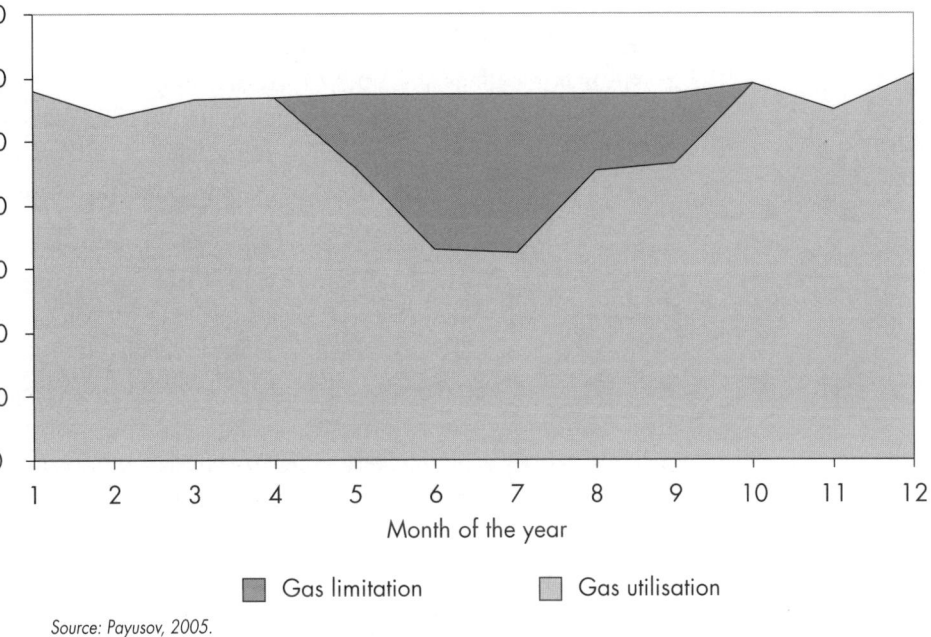

Source: Payusov, 2005.

Figure 26 shows the limitations of Gazprom's network access to the associated gas of Yukos at its Tomskneft fields due to seasonal factors (from April to October). This highlights the problems limiting the utilisation of associated gas resources in Russia from the stated lack of capacity of the transmission network, the lack of storage capacity and the failure to co-ordinate long-term development programmes between oil companies and Sibur.

In this respect, investments such as those made by Yukos to use 2-3 bcm of associated gas per year to fuel co-generation plants at an estimated cost of USD 200 million, provide an economic alternative for Russian oil companies. It also avoids the issue of unreliable access to Gazprom's pipeline. An oil company's ability to use the volumes of associated gas it produces is hampered by the limited need for gas to generate electricity for its own use as well as the limited gas needs of the local community unless the oil field is located relatively close to a large urban centre. There are only a few examples where oil companies have invested in compressor stations and connecting pipelines to the Gazprom transmission system in order to supply gas to regional consumers (Box 12). However, the potential to replicate these types of projects is limited by Gazprom's transmission capacity in each particular part of the system.

Lukoil

Lukoil, Russia's largest oil producer, accounts for about 10% of the associated gas produced in Russia per year and about 10% of the officially reported volumes of flared gas. Since 2002, Lukoil has made efforts to increase its utilisation rate of associated gas, reporting an increase from 74 to 80% in 2004, but with a drop to 77.5% in 2005. Lukoil supplied about 2.8 bcm of its associated gas to Gazprom's Sibur gas processing plants in 2004. A further 1.3 bcm was used in power generation for its own use and another 0.3 bcm was used for local consumption. The remaining

Box 12 Tomsk gas flaring reduction project

In June 2002, the Yukos subsidiary, Tomskneft, commissioned a gas compressor station at the Luginetsk oil and gas condensate field in the Tomsk region. In late 2003, Yukos attributed a 15% reduction of associated gas flaring (equivalent to 85 000 tCO$_2$) to this investment. The gas compressor station has a capacity of 1.5 bcm per year, allowing Yukos to increase both oil and natural gas production at nearby fields, some with high gas-to-oil ratios. The processed natural gas enters Gazprom's transmission system via a pipeline built by Yukos. Industry and electricity plants are major consumers of this gas as are West Siberian and Altai cities connected to Gazprom pipelines.

Source: Yukos, 2003.

1.1 bcm was flared out of a total of 5.6 bcm of produced associated gas. In 2005, 1.4 bcm was reportedly flared. Lukoil considers installing gas turbine power generators close to oil fields as an efficient use for associated gas, given it reduces flaring, as well as electricity and production costs.

In mid-2003, Lukoil signed a long-term contract with Gazprom to purchase its associated gas for USD 22/thousand m^3. It covers the supplies of gas produced by Lukoil at the Nakhodkinskoye field in Yamalo-Nenetskiy Autonomous Region. This contract was concluded between Lukoil and Gazprom in the General Agreement on Strategic Partnership for 2002-05. Although much lower than export market prices and lower than domestic gas prices, it provided Lukoil with some minimal monetisation for this production stream and avoided costs associated with flaring. This is especially important after the Russian government increased the fees for methane emissions and gas flaring in mid-2005 (see Chapter 2).

Over 2006, Lukoil plans to increase its natural and associated gas production to levels of 8-10 bcm/y from just over 6 bcm in 2005. Over the next 5-10 years, Lukoil foresees expanding this to production levels in the order of 50 bcm/y. The question for Lukoil is not how to produce the gas but what to do with it once produced. Lukoil is looking at various regional markets as well as gas use for electricity generation or in petrochemical processes. Like other Russian oil majors, Lukoil is interested in positioning itself to benefit from improvements in the economics of gas sales on the Russian market as domestic gas prices rise (Weiss, 2006).

Satellite imagery calibration methodology: a new tool to monitor and stimulate compliance?

Unofficial estimates place the volume of gas flared in Russia much higher than official figures – in some cases more than double. According to World Bank conservative estimates, Russian volumes accounted for about 10% of the world's flared gas in 2004.[160] There are some obvious difficulties with data transparency and consistency in associated gas flaring statistics worldwide. Russia is no exception. Furthermore, in Russia, this data is reported to several governmental agencies and committees at the federal and regional level without coordinating to cross-check the information

---

160. This corresponds to the officially reported data of the GGFR.

(GGFR, 2004a). Lack of meters and limited monitoring capacity is yet another key factor contributing to possible inconsistencies in officially reported statistics.

A simple calculation can be performed to estimate the maximum amount of associated gas which can be technically flared in West Siberia on an annual basis, given the following information and assuming all associated gas was flared:

■ Russian oil production in the order of 9 million barrels per day two thirds of which from West Siberia.

■ A maximum GOR estimated at 200 m^3/m^3 in comparison to the average GOR of 100 m^3/m^3 (based on US EIA (1997) and *Energy Sector of Russia* (2006)).

Based on these assumptions, one can estimate the maximum volume of associated gas that can be technically flared in Russia in the order of 110 bcm per year. This means 75 bcm per year is the maximum volume for West Siberia.

The IEA in co-operation with the National Oceanic and Atmospheric Administration of the United States (NOAA) undertook a preliminary study to estimate the volume of gas flared in West Siberia using satellite imagery. Map 2 presents a composite of clear-night images of West Siberia from the Defense Meteorological Satellite Programme (DMSP) satellite F-16 for the year 2004. Another satellite image was used as a "sample" covering a site where the annual rate of flared volumes was known, to calibrate against the flares evident in the satellite image of West Siberia.

Ideally more metered data points at more sites are necessary to enhance accuracy of the calibration. However, there is little such data. Countries and oil companies with good quality monitoring tend to have very few continuous flares, while countries where flaring is routine tend to have limited monitoring capacity – or little interest in monitoring.

The black areas on Map 2 represent points of light including flares which had lighting 90-100% of the time every night in 2004 with intensity at least equal to that of the sample flare.[161] The white crosses represent the reported geographic location of the West Siberian oil fields. As a first step, we assume all the black areas are gas flares, understanding however, that this will need to be refined to eliminate other sources of heat such as city lights or industrial activities. With the sample flare, we estimate the volume of gas flared in West Siberia using two different approaches.

Estimate based on the area of lighting

The volume of gas flared can be estimated based on the area of lighting produced. The sample field flare produced 42 km^2 of lighting with a digital number (DN) of 30 or more. The DN is a measure of light intensity on the image. Based on the known rate of flaring from the sample flare of 0.85 Mm^3/day of flared associated gas, the sample flare shows 0.02 Mm^3 per km^2 of lighting with a DN greater than 30.

---

161. DMSP F-16 night-time lights straddle the visible and near infrared from 0.5 to 0.9 μm.

Map 2

Satellite image of West Siberia

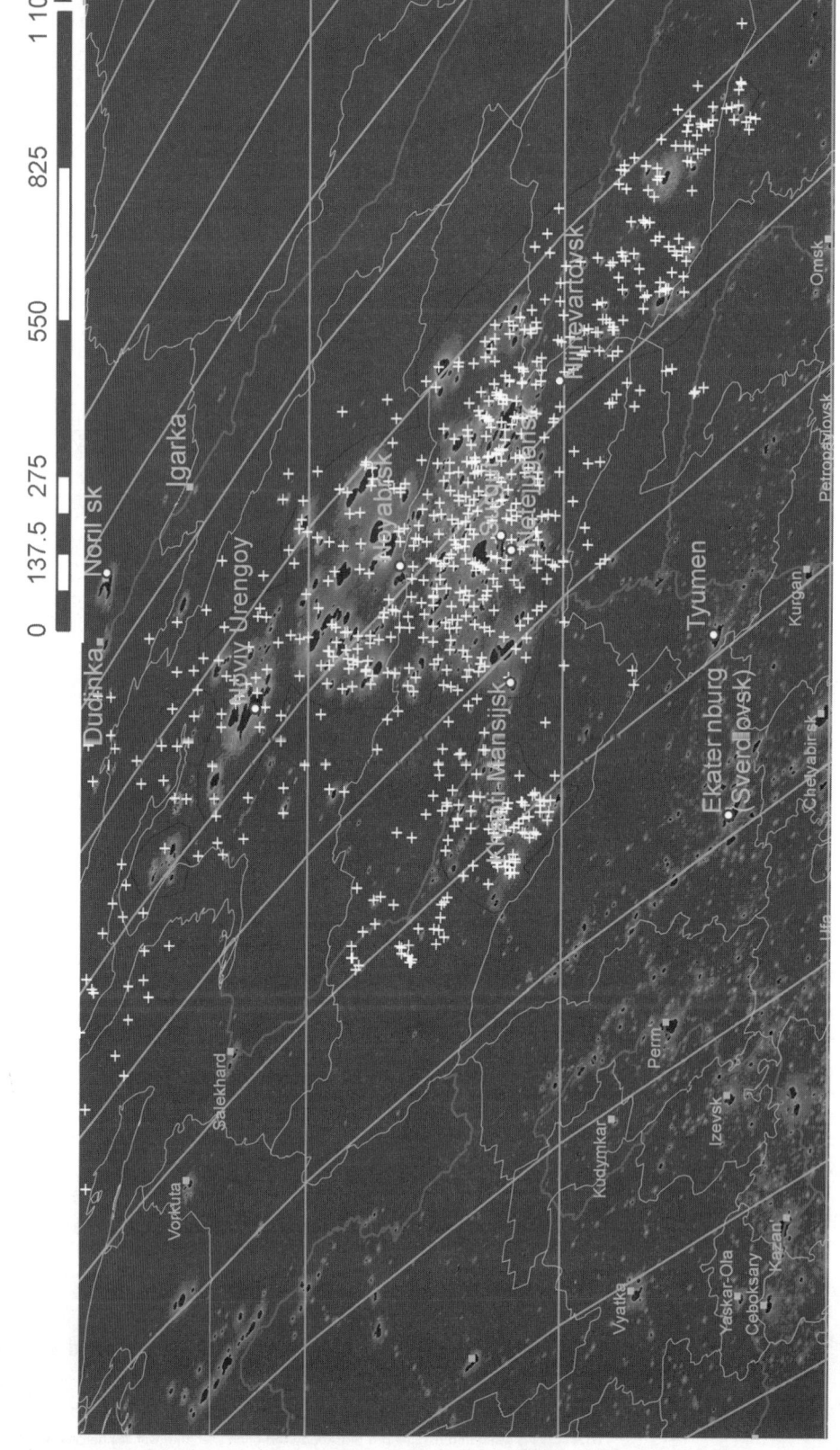

Source: US NOAA, 2005; US EIA information, 1997.

The West Siberian image has 27 812 km^2 of lighting with DN greater than 30. This gives an estimate of 27 812 x 0.02 = 562 Mm^3/day. On an annual basis this represents 205 bcm/year. Clearly this is an overestimation in comparison to the maximum volume of gas that could technically be flared in West Siberia. For this reason, an alternative methodology was adopted to try to eliminate the lighting points clearly not related to the flaring of associated gas.

Estimate based on intensity and size of the sample flare

An alternative methodology is to tally the flares in West Siberia with an intensity and size larger than those of the sample flare. This raises the error margin due to double-counting of overlaps between lighted areas. Using this methodology, 328 flares or lighting points can be identified in the West Siberian region. All the black areas have a DN greater than 30. Thus 328 x 0.85 Mm^3 = 101 bcm/y.

Several reasons explain our possible overestimation of volumes of gas flared. Examination of the satellite photo from the sample flare site reflects city lights and other industrial installations. One can distinguish flare points and city lights with the aid of a map and the knowledge of where large oil and gas activity is or is not being conducted. Thus "apparent flares" can be clearly distinguished between city lights or gas flares.

Among the big "apparent flares" in the satellite image of West Siberia, lighting points corresponding to cities such as Sverdlovsk, Tyumen, Surgut, Nizhnevartovsk, Nefteyugansk, Khanty-Mansiysk, Norilsk and others can be eliminated from the calculation. Furthermore, West Siberia is an important industrial region with about 10 gas processing plants, refineries and huge mineral smelting plants (including the Norilsk nickel plant). According to our preliminary estimates these "apparent flares" represent about 15 to 20% of the lighting points on the satellite image in Map 2. This reduces our estimates of volumes of flared gas to about 60 bcm.

Overestimations could also be due to "technological" factors, such as burner design and the local environment, which affect the heat radiation from flares. The calibration from light intensity to volume of gas flared may be inconsistent due to:

■ Burner designs which affect flame temperature, size of flame, soot formation, and presence of liquid droplets. These factors can affect significantly the emissivity of radiation. Flame temperatures, for instance, can vary by 20% depending on burner design. This could affect the total radiation by a factor of 2 or 3. In terms of the satellite data this could be amplified 4-5 times (*i.e.* by a factor of 10) given the focus on the near infra-red part of the light spectrum and the near infra-red emission.[162]

■ Atmospheric humidity which can also play an important role in changing the rate of light absorption along the path between flare and satellite. In this respect, the atmospheric humidity above the sample flare site may not be the same as that in West Siberia.

■ Reflectivity (from the sea or snow cover) which could also add to the uncertainty.

162. This is estimated assuming a blackbody radiation spectrum with temperatures in the order of 2 000° Kelvin.

Next steps and recommendations

These difficulties serve to illustrate the point that more detailed flaring data from oil companies active in West Siberia is necessary to achieve a clear picture of the extent of flaring. Although burner designs are most likely to be different in Siberia than in other parts of the world, they are more or less standardised within Siberia. In this respect, the work of the GGFR and its co-operation with the authorities of the Khanti-Mansiysk Autonomous Region could prove extremely useful. At the federal level, the Russian government's increase in CH_4 emission fees may also provide more political support for this new methodology as the government will need improved monitoring capabilities to ensure effective implementation.

Russia's decision to focus on energy security during its G8 Presidency in 2006 could provide a useful platform on which to raise the profile of this new methodology. The Russian government could support this initiative as a way to provide a model for other major gas-flaring regions of the world. This would provide a useful step forward in tackling this issue – one that will raise transparency, enhance energy security and reduce the negative impact of gas flaring on the environment globally.

Regulatory requirements to reduce gas flaring

The regulatory requirements governing associated gas utilisation and/or flaring in Russia are included in the licenses, issued jointly by the Ministry of Natural Resources and regional authorities.[163] According to the license, operators may lift, process, and market associated gas; use it in operations, re-inject it or flare specified volumes.

The Federal Mineral Resource Act (1992) does not set any limitations on associated gas flaring and usage. Currently, no specific secondary legislation (such as codes, guidelines) exists at the federal level to deal with operational processes or regulatory procedures related to gas flaring or venting. Restrictions on flaring are region-specific and depend on regional policies on this issue.[164] Only a few regions have included special provisions on associated gas in their regional Mineral Act.

The Khanti-Mansiysk region accounts for close to 60% of Russian oil production and two-thirds of Russia's associated gas production. About 20% of its associated gas is still flared, representing 40% of total gas flared in Russia (6 bcm officially reported in 2004). Khanti-Mansiysk region includes the usage rate of associated gas as a mandatory license condition. It has capped gas flaring at a level of 5% of gas produced by license operators. The rate is negotiated for each license between the operator and the authorities and can be relaxed if the operator demonstrates that this level is unrealistic or unattainable. In 2005, the administration of the Khanti-Mansiysk region reported only about 30% of licences was meeting the required 95% associated gas utilisation rate (see Table 32).

According to the GGFR (2004a), "oil companies often opt not to negotiate higher gas-flaring limits, since compliance is unlikely to be scrupulously monitored". Russian authorities lack the capacity to enforce these limits. In theory, the supervision

---

163. For more detail see the GGFR's Russian regulatory review (GGFR, 2004a).
164. Relevant legislation on gas flaring and venting in Russia is listed in GGFR (2004a).

Table 32   Compliance with licence requirements on rates of associated gas utilisation in the Khanti-Mansiysk region

Oil company	Number of licences	Compliance with licence requirements on rates of associated gas utilisation		
		Complete compliance	5% non-compliance	Non-compliance
Lukoil - West Siberia	44	4	15	25
Surgutneftegas	34	18	1	15
TNK-BP	32	7	4	21
Yugaskneftegaz	26	9	8	9
Slavneft-Megionneftegaz	12	2	1	9
Russneft	9	1	0	8
Tomskneft	7	4	0	3
Sibneft	5	0	1	4
Other oil companies	44	10	4	30
Total in the region	213	55	34	124
Rate of compliance	**100%**	**26%**	**16%**	**58%**

Source: Khanti-Mansiysk administration, 2005.

of gas flaring is the responsibility of the Ministry of Natural Resources and regional authorities. While non-compliance of gas-flaring limits can be just cause for license withdrawal this has never been done to date.

The GGFR (2004a) highlights that the effective enforcement of gas-flaring limits is hampered by the lack of clarity in defining the division of responsibilities between authorities and supervisory agencies at the federal and regional levels, as well as by the lack of standardised reporting, monitoring, and enforcement procedures for oil companies.

Proposals of oil companies

Lukoil is advocating the complete prohibition of gas flaring at new oil fields as a way to focus attention on the need to establish a comprehensive unified technological system from producing associated gas to processing it into products with high-profit margins – for use in local electrification, other energy needs (large fraction of light hydrocarbons and LPG) or in the petrochemical industry (Astakhov, 2005). Lukoil also stresses that the Russian government should introduce sanctions against the flaring of associated gas and establish priority access for associated gas to Gazprom's pipeline system.

Lukoil has stated that more should be done at the government level to provide incentives to reduce gas flaring such as:

■ Changing the Subsoil Law to ensure associated gas utilisation levels no lower than 95%.

■ Increasing consumer product prices.

■ Introducing fees for gas flaring on the basis of cubic meter as opposed to concentration of harmful emissions.

In 2005, Yukos also supported the necessity for legislative changes to overcome the following deficiencies:

■ The disconnect between the Subsoil Law which obliges the subsoil user to meet the technical requirements set out in its work programme (*i.e.* 95% utilisation), and the Tax Code which sets the tax rate based on 100% utilisation of associated gas.

■ The fact that associated gas is not included in the list of resources for which normative losses are established (Russian Government, 2001a).

■ The contradiction within the Law "On Gas Supply" to uphold the state regulation of rational use of gas reserves and to ensure access to Gazprom's transport system by independent organisations while at the same time providing a legal basis for Gazprom to limit this access. The Law states (Russian Government, 2001b): "Gazprom shall ensure access to independent organisations to its gas transportation system on the basis of agreements… on the condition there is excess spare capacity in the gas transportation system from the point of intake to the point of delivery for the proposed gas supplier for the required period".

Regional proposals: Khanti-Mansiysk region

The Khanti-Mansiysk region has been a partner of the GGFR since 2004 and has worked actively on encouraging the use of associated gas in its oil fields. The regional administration has drawn up a set of measures, which could be implemented at the federal and regional level to resolve the issue of gas flaring. These proposals are in parallel with those recommended by Russian oil companies, setting out an efficient framework that could lead to a win-win situation for private and state interests. Khanti-Mansiysk proposes possible national measures similar to those suggested by the GGFR "Voluntary Standard for Global Gas Flaring and Venting Reduction" (see Table 33).

The main measures proposed to be implemented at the federal level include (Khanti-Mansiysk Administration, 2005):

■ Development of specific legislation "On associated gas" establishing the rights and responsibilities of the authorities, oil producers, infrastructure owners and gas processing plants to meet the mandatory gas-flaring limits. This would provide a

Table 33   GGFR Voluntary Standard options for governments

GGFR Voluntary Standard options for consideration by governments
Clarify regulatory framework and contract rights
Promote third party access to infrastructure
Consider Production Sharing Agreements (PSA) allowing cost recovery of gas infrastructure
Examine cost recovery and profit-sharing mechanisms
Institute tax and royalty incentives for gas use
Improve the pricing of associated gas
Develop a national gas strategy including associated gas
Develop local markets improving legal and fiscal frameworks for large consumers (power sector)
Provide payment guarantees to reduce the financial risk of producers
Co-ordinate stakeholders to enhance opportunities for gas utilisation

Source: GGFR, 2004b.

legal recourse for oil companies which would otherwise have to assume the entire cost of penalties (*i.e.* emission fees) if forced to flare due to limited access to pipelines and gas processing facilities.

■ Establishment of incentive mechanisms, such as tax or royalty abatements, for operators using new technologies and equipment to increase the use of associated gas. These incentives are of particular relevance for operators of new fields or fields with unfavourable conditions for gas utilisation.

■ Development of a long-term federal programme on the utilisation of associated gas (15-20 years), including region-specific provisions. This programme should encompass the development of gas processing capacities to ensure the utilisation of associated gas from oil companies and gas from independent gas producers.

■ Establishment of a mandatory 5% cap for gas flaring at each oil field.

The Khanti-Mansiysk regional administration points to the need for structural change in the gas sector in order to resolve the problem of third party access to Gazprom's transmission pipelines and its gas processing facilities. It also proposes more cost-reflective regulated gas prices.

The region is also working on establishing common guidelines for measuring and reporting associated gas utilisation and flaring, to improve and standardise monitoring procedures. Best measurement practice requires continuous metering at the source or at the flare burners, to determine the annual volumetric flow of the flare (GGFR, 2004c).[165] The GGFR "Voluntary Standard for Global Gas Flaring and Venting Reduction" recommends this practice to be used for all new projects and existing large projects. For Kyoto-related projects, the use of best measurement practices may also be chosen to calculate emission reductions. However, at existing fields the installation of metering equipment is not always possible or economic. Instead, volumes of gas flared and vented could be estimated based on mass and energy balances.

Federal proposals   Until recently, the Russian Federal government has focussed little attention on gas flaring. However, in July 2005, the government agreed to increase the existing fees for CH_4 emissions by a factor of 1 000, albeit from a low base (Russian Government, 2005). These fees target methane emissions from leaks of equipment and components of natural gas systems and also the methane contained in associated gas flared by oil companies.

The effectiveness of this federal government initiative depends on whether monitoring agencies are able to implement and enforce these new fees. Russia's regulatory bodies often lack the financial and human resources to carry out effectively the tasks set out in environment regulations and legislation (MEDT, 2006b). This is especially a concern when these regulatory bodies face the challenges of unequal access to information given the difference in scale and resources between the regulator and the huge companies they are expected to monitor and regulate.

---

165. Flow-measurement devices to determine flare volumes have an accuracy of ±5%.

IMPACT OF CARBON FINANCE

Kyoto-related mechanisms may provide additional economic incentives to reduce gas flaring in Russia. In this section, we assess whether a value stream for CO_2 emission reductions significantly alters the economics of projects to enhance utilisation of associated gas. Can the carbon price be a strong enough incentive to increase the number of gas flaring reduction projects in the current regulatory and economic environment in Russia? Bearing in mind that the answer will be project specific, our assessment is based on an "average" oil production area roughly representing a typical West Siberian oil producing field.

Given that at the initial stage of Kyoto-related activities in Russia, oil companies will have access only to JI Track 2 carbon financing, additionality for gas flaring reduction projects is briefly discussed. The additionality demonstration procedure could be facilitated in the framework of JI Track 1 or GIS if the Russian Designated National Authority (DNA) develops more comprehensive national guidelines that will take into account the specific sectoral circumstances.[166]

Analysis of a "typical" oil field in West Siberia

In our analysis, a production area is considered: a set of fields that are in relative proximity to each other which jointly produce a significant amount of oil and associated gas. Examples of such groupings could be all the fields near Nefteyugansk (*i.e.* the Yugansk production unit of Rosneft), or all the fields near Surgut (*i.e.* a large part of the production of Surgutneftegas). We assume a daily oil production of 1 Mbbl per day with a GOR of 100 m^3/m^3, an average for the West Siberian fields. This means that 16 Mm^3 per day of associated gas is produced. Our average production area is located close to a city of 100 000 people (typical of West Siberian "oil towns"), but otherwise 1 000 km away from any large industrial centre and 500 km away from access to existing Gazprom gas transmission pipelines.

For this "typical" oil field, the attractiveness of three possible options for utilisation of associated gas is considered for the volumes left after local needs:

■ Pipeline transport, which as shown in Figure 29, is likely to be the preferred technology for the conditions we assumed.

■ Gas-to-liquids (GTL) option is explored as an alternative to the limited and uncertain access to Gazprom's pipelines. The GTL option is relatively immature, risky and costly, with only a few plants currently operating or being built worldwide. For this reason, this cannot be considered as a near-term solution. Major multinational oil companies, as well as some Russian oil companies and Gazprom are actively engaged in R&D of GTL, suggesting that this may be an attractive possibility in the future.[167]

■ Re-injection of associated gas in oil fields to improve recovery of oil.

---

166. The possibility to use gas flaring reduction projects to generate emission reduction units could be even more straightforward if the upstream activities of oil companies could be included in a domestic ETS.
167. Gazprom and Yukos are working with an American developer on assessing potential GTL sites in Russia.

Local use of
associated gas

Use in the oil-extraction process. Most of the wells in West Siberia use down-hole electrical pumps and produce at an 80% water cut (1 barrel of oil for every 5 barrels of fluid produced). One can estimate that about 1 kWh of electricity is needed to lift 1 barrel of fluid with 2 000 m head with 75% efficiency. Thus, each barrel of oil produced requires about 5 kWh of energy. We consider, as a conservative estimate, that total energy needs of oil production, taking into account all other energy use (*i.e.* drilling), is about 10 kWh per barrel. If this energy is produced from associated gas using a low-efficiency gas turbine with 30% efficiency, an estimated 30 kWh per barrel will be needed. Given the gross heat content of gas in Russia of 38.231 MJ/m^3, the daily production of associated gas (16 Mm^3) translates into 160 kWh. The above assessment shows that only about 20% of associated gas would be needed in the oil extraction process.

Use to supply energy needs of local cities. The energy consumption by a local city with a population of 100 000 can be estimated at 1 200 toe/day based on the average per capita primary energy demand for Russia of 4.3 toe.[168] On this basis, only about 10% of associated gas would be required by a local city.

Thus, 30% of the volume of associated gas production would be sufficient to meet the energy requirements of the oil extraction process and for local city needs.

Long-distance
transport to
a market: the
pipeline option

Economic assumptions

Pipeline investments (Capex). If an investor were to build a pipeline to connect to the main transmission line, the following components would be necessary:

- A gas treatment plant to ensure a gas composition suitable for access to a Gazprom pipeline.

- Two compressor stations to bring the gas from essentially atmospheric pressure to the standard pressure of Gazprom's long-distance pipelines, and to maintain this pressure along the transportation distance.

- Approximately 500 km of pipeline to link the production area to the nearest access point along Gazprom's transmission system.

Estimates of pipeline costs based on expert advice and engineering journals range from USD 0.5 to 1 million per km including compressor costs. Based on this, we estimate capital costs in the order of USD 700 million for a throughput of 10 Mm^3 per day, assuming the rest is consumed locally.

Pipeline operating costs (Opex). Pipeline operating costs consist mainly of maintenance and compression costs. For a 500 km pipeline, Opex is estimated at USD 0.01/m^3. In addition to the cost of standard pipelines, for associated gas there is also the cost of initial compression from essentially atmospheric pressure to 75 atm.

---

168. Per capita primary energy demand in Russia was 4.3 toe per year in 2003. This is an upper estimate given the average per capita demand includes energy used by industry (IEA, 2004b).

This can be estimated at about 0.5 kWh/m^3. As this can be provided using about 5% of the associated gas, we do not include this in our calculation.

Revenues from gas sales. Given the current market structure in Russia, we assume associated gas will be sold on the domestic market at the Russian domestic price of USD 40/thousand m^3. However, Gazprom has no reason to pay more than its marginal cost of production, which is estimated in the order of USD 22/thousand m^3.[169] However, given the decline in production of existing fields requiring investment in new higher-cost fields, Gazprom may soon be ready to pay significantly more. Russian domestic gas prices in 2010 are projected to reach USD 60/thousand m^3.

GHG emission reductions. A conservative estimate of the volume of GHG emissions due to the flaring of associated gas is based on the emission factor 2.9 kgCO_2/m^3 (see Annex 1). The pipeline and compressor stations used for the transportation of associated gas will induce CH_4 leaks from equipment during operations and maintenance and the energy-related CO_2 emissions from compressors. Therefore, we assume that only about 75% of GHG emissions can be reduced, *i.e.* 2.2 kgCO_2e per m^3 of gas.

At current domestic gas prices and based on the assumptions of our "typical" example, it is not economically attractive to use associated gas if a pipeline needs to be built (see Figure 27). Of course, the entire concept is vulnerable to the uncertainty of access to Gazprom's transmission pipelines.

To evaluate the impact of carbon revenues on the economics of this option, we assume that the investor obtains the emission reduction credits for the whole amount of projects' emission reductions, without considering the issue of additionality. Figure 27 shows the estimates of internal rate of return (IRR) of a pipeline project (over a 10 year period) as a function of the associated gas price with and without carbon revenues derived from sales of ERUs at a price of USD 7/tCO_2.

The carbon revenue stream has a significant impact on project economics, especially given current low domestic gas prices (IRR increases by more than 10%). This rough analysis suggests that there will be projects (for instance with shorter pipelines, lower compressor costs, higher flow rates or volumes of associated gas) for which the ERU revenue will enhance project economics enough to attract investment. In this case, reliable access to Gazprom infrastructure will be critical in the investment-decision process.

In terms of the use of Kyoto Protocol flexibility mechanisms, additionality will be a key issue. For example, one could argue that the amount of additional ERUs and therefore the carbon revenues would be considerably lower if the baseline was based on the license terms which stipulate 95% use of associated gas. This more strict approach to additionality would mean that investors could benefit from only 5% of total GHG emission reductions.

---

169. Based on the 2003 agreement between Gazprom and Lukoil.

Figure 27   Impact of the associated gas price and carbon revenue on the IRR of pipeline-transport projects

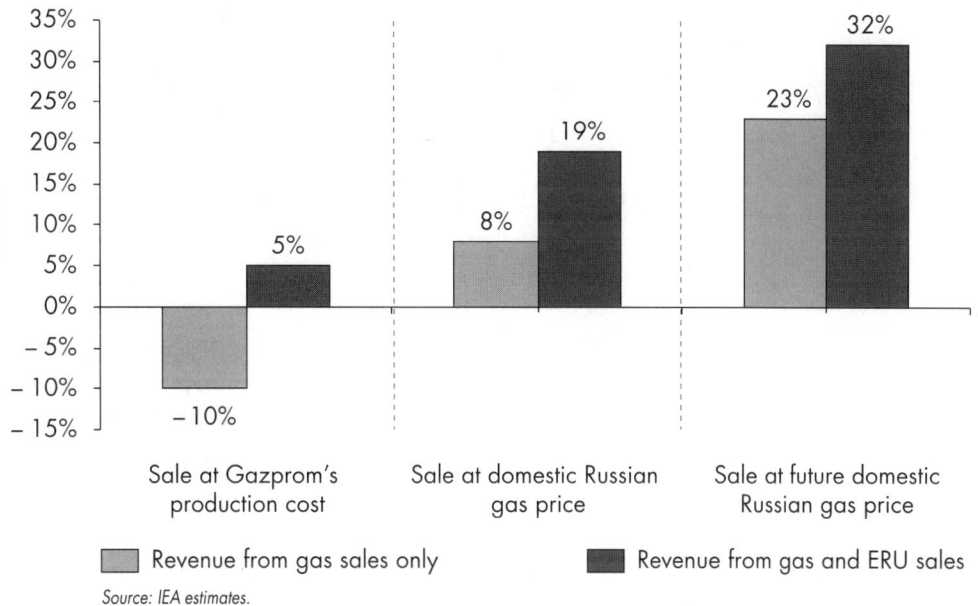

Source: IEA estimates.

The analysis shows that at higher domestic gas prices, projects including the construction of a connecting pipeline to the Gazprom system, would be attractive. The internal rate of return for this "typical" pipeline project is close to 25% even in the business-as-usual condition without carbon revenues. However, under current regulatory conditions, even if the domestic price of gas were attractive enough to support such projects, the ultimate barrier would be the absence of reliable, fair and transparent third party access to Gazprom pipelines.

Long distance transport to market: gas-to-liquids option

In the event that Gazprom does not provide reliable third party access in the longer-term, an oil company may find it attractive to look at other options such as gas-to-liquids (GTL) using its existing oil transport infrastructure and liquid hydrocarbon markets.

Capex of a typical GTL plant. Estimates range from USD 20 000 to 40 000 per barrel of liquid capacity per day. We assume an average cost of USD 30 000 per day. Process yield is typically 1 barrel of liquid for 283 m³ of gas. For the processing of 10 Mm³ per day of associated gas, Capex is in the order of USD 1 billion.

Opex of a typical GTL process. Estimates range from USD 4 to 10 per barrel of produced liquid. With associated gas used as input gas (instead of being flared), it is considered a free input.

Revenue from sales of liquids. GTL normally produces diesel-like middle distillates that carry a price premium of USD 4 to USD 6 per barrel over crude market prices. Whether this can be realised depends on the existing liquid transport infrastructure. It may turn out that the only economic approach is to mix the GTL liquids with crude oil in oil pipelines, thereby losing most of the premium. We assume a long-term

Brent price of USD 25/barrel (a conservative assumption used by oil companies as a benchmark for their investment decision making) and the cost of West Siberian crude at the wellhead from USD 10 to 15 per barrel. Taking into account the premium and the Opex, we estimate a net revenue from USD 5 to 15 per barrel of synthetic liquid fuel. The mid-range value of USD 10/barrel is assumed, giving a net revenue of USD 0.035 per m^3 of gas.

GHG emission reductions. Based on the process efficiency from existing international projects, we assume that 70% of the GHG emissions from associated gas flaring is eliminated using the GTL option (2.0 kgCO$_2$e/m^3).

In the business-as-usual case, the GTL option is not economically attractive (IRR=3% over a 10-year period). The additional carbon revenue at USD 7/tCO$_2$ significantly improves project economics, increasing the IRR by 8%. However, to overcome an investment hurdle rate of 25%, project developers will need an ERU value of at least USD 22/tCO$_2$.[170]

Gas re-injection option

In our "typical" example, the geological conditions are assumed to be favorable for gas re-injection in the oil reservoir.[171] Both reservoir and aquifer injection are assumed to have identical costs. The additional reserves or future gas production are not taken into account. Moreover, the aquifer pressure is assumed to have the same order of magnitude as the reservoir pressure.

Capex of gas re-injection includes the cost of a gas treatment plant (for example to remove liquids to improve injectivity), pipelines between producing wells, compressors, and injection wells. To simplify, only compressors, which usually represent the largest cost, are taken into account. We assume that wells that can be used for injection already exist, as well as the pipeline infrastructure, and that no treatment plant is needed.

To provide an approximate estimate of compressor costs, we use the data published by Statoil for CO$_2$ re-injection in the Sleipner and Snøhvit projects, _i.e._ USD 80 million and USD 70 million for 1 MtCO$_2$/y and 0.7 MtCO$_2$/y, respectively. We assume similar annual costs for gas re-injection. Given these projects are offshore using western compressors, typically more expensive than Russian ones, we have halved the costs to an estimated annual cost of USD 0.1 per cubic meter of re-injected gas. Assuming a capacity of 10 Mm^3/day, capital costs are estimated at USD 360 million.

Opex of gas re-injection includes well and compressor maintenance costs, and the cost of energy used by the compressors. Although energy input would typically dominate the operating cost structure, given the use of associated gas which would otherwise be flared, we can assume Opex for the gas re-injection project in the order of only USD 0.005 per m^3 re-injected.

---

170. This supposes also that all generated emission reduction units (ERUs) are credited.
171. The technical feasibility of this option cannot be estimated without a detailed study of a specific reservoir.

GHG emission reductions: the estimates of energy required to compress the gas suggests that 10% of the gas will be used for compressors. Indeed, about 90% of the current emissions from flaring can be reduced, *i.e.* 2.5 $kgCO_2$ per m^3 of associated gas.

If one does not take into account the amount of avoided fees for methane emissions, the emission reduction credits are the only short-term revenue stream of a gas re-injection project. With the lower-bound ERU price (USD 7/tCO_2), which can be expected in high-risk host countries, the carbon revenue would not be sufficient to raise the internal rate of return above the investment hurdle rate (IRR=5%). If the ERU price is in the upper bound level of USD 14/tCO_2, the project is much more attractive (IRR=28%).

Conclusions on "typical" projects to enhance the use of associated gas

Our assessment of "typical" projects illustrates the reasons why gas is flared in Russia. Current domestic gas prices and unreliable third party access to Gazprom's infrastructure are the two key reasons. Furthermore, the current ERU prices are not able to raise project economics to viable levels. Higher ERU prices, reflecting to some extent lower host-country risk in terms of Kyoto-related projects, have a significant impact on project economics. In the case of the gas re-injection option, as well as for the technologically immature GTL option, carbon finance could contribute to an attractive return on investment, and could drive investment decisions. This is particularly true for the gas re-injection option, for which the sale of the ERUs would be in most cases the only direct revenue stream.

The Russian government's attempt to stimulate projects to reduce gas flaring through an increase in fees for CH_4 emissions (discussed in Chapter 2), may raise the cost of gas flaring in Russia and thereby provide more incentives for companies to find alternative uses for the gas. However, the effectiveness of such administrative regulation will depend on the monitoring capacity of authorities to enforce the collection of environmental payments.

Access to carbon finance: the additionality issue

Currently, Russian oil companies may have access to carbon financing only through the JI Track 2 process. Thus, these projects will have to demonstrate the *additionality* of their GHG emission reductions in comparison with the baseline scenario (see Box 4).

The methodological rules for JI Track 2 are not yet established. However, as discussed in Chapter 2, the relevant CDM experience may provide some insights on possible future JI Track 2 guidelines and the approved CDM methodologies can be used for JI Track 2 projects. In this case, the experience of the CDM Rang Dong project in Vietnam (see Box 13) may be useful to understand the possible ways gas flaring projects could be accepted under JI Track 2.

The GGFR has developed guidelines for the baseline methodology and the demonstration of *additionality*, which take into account the specificity of CDM gas flaring reduction projects (see Table 34).

Box 13   The Rang Dong project of recovery and utilisation of associated gas

The Rang Dong project includes construction of a gas pipeline and compressor facilities to recover and transport associated gas, which would otherwise have been flared. The Rang Dong oil field is located about 140 km off the south-eastern coast of Vietnam. The project has resulted in the reduction of gas flaring. The recovered gas is processed into dry gas (mostly methane), as well as LPG, and condensate (hydrocarbons with $+C_5$). The dry gas is to be supplied to the nearby power plants, whereas LPG and condensate will be consumed domestically as home cooking fuel and octane enhancer of gasoline respectively. The net volume of CO_2 emissions eliminated is approximately 6.74 $MtCO_2$ over the 10-year crediting period.

To determine the baseline scenario for the Rang Dong project, the following alternatives for associated gas utilisation were assessed:

■ Release into the atmosphere was not considered as a possibility given that it is prohibited by law.

■ Gas re-injection into the oil reservoir is highly unattractive compared to water injection.

■ Construction of a pipeline is unattractive in the absence of emission reduction credits (IRR only 8-9%) considering the cost and low revenue stream from gas sales to power plants.

■ Local consumption.

■ Gas flaring.

The two last options are not prohibited by regulation and represent current practice in Vietnam. Thus, the project baseline is a continuation of current practices with the partial use of gas for on-site electricity while the remainder of the associated gas is flared. In the Rang Dong case, additionality is demonstrated on the basis of investment analysis. The baseline study shows that the construction of a pipeline is not attractive in the business-as-usual case. This option becomes economically feasible only in the framework of CDM when carbon revenues from the crediting of GHG emission reductions are possible.

Source: CDM EB, 2005c.

According to this methodology, the assessment of a country's regulatory requirements is a first step in determining possible gas flaring reduction options. If flaring is banned by regulation, then it cannot be considered as a baseline scenario and its reduction is not considered additional. This approach is appropriate in countries where strict enforcement of regulatory requirements for associated-gas use is common practice.

Table 34   Additionality demonstration tools for gas flaring reduction projects

Narrowing baseline options

Plausible and permissible associated gas development options:
    Regulatory requirements
    Technological feasibility
    Geophysical conditions

Qualitative or quantitative assessment
Green-field projects: compare attractiveness of investment options Brown-field: compare investment against "business-as-usual" based on a set of relevant economic and financial indicators
Barrier analysis
Domestic regulated price of associated gas, domestic fuel prices, risks related to local markets Fiscal regimes, Production Sharing Agreements, and other regulatory risks Technological risks (new technology) Implementation risks
Reference to *common practice*
Demostrate empirical evidence on *common practice* Use other tools if *common practice* demonstration is not relevant
The sequence of tools may differ depending on projects.

Source: GGFR, 2005b.

For countries where legal requirements are ambiguous or not enforced or exemptions are granted, this type of approach may lead to the disqualification of projects that might otherwise provide real sustainable environmental benefits in terms of GHG reductions.

As illustrated by our assessment of options to enhance the use of associated gas in Russia, the current structural and market barriers of the gas sector can render gas flaring reduction projects unattractive for oil companies. These barriers should be taken into account when determining the additionality of gas flaring reduction projects, if the quantitative assessment (using economic and financial indicators) qualifies the project as *a priori* feasible. In this case, some projects to enhance the use of associated gas may be considered additional even if they aim to comply with the mandatory limits established in licenses, given that this would otherwise be impossible to achieve in the current regulatory and pricing environment in Russia's business-as-usual case.

As discussed in Chapter 2, in the context of JI Track 1 and/or GIS, the demonstration of a project's additionality may become more straightforward if the DNA were to put in place comprehensive guidelines clarifying rules to take into account the national and/or sectoral policies and circumstances in a baseline scenario.

Carbon revenue could become one of the incentives to speed up the pace of implementing gas flaring reduction investments in Russia. However, the use of alternatives to gas flaring is limited by the unreliable access to Gazprom pipeline infrastructure and low domestic gas prices. Given these fundamental barriers related to lack of progress in reforming the market and energy price structure, Kyoto-related mechanisms alone can not provide a unique tool to resolve gas flaring problems in Russia. As our assessment of the "typical" options shows, significant improvements in the regulatory framework and market structure of Russia's gas industry will be necessary if Russia is to effectively address the problem of gas flaring in a sustainable way.

Just as the Russian Energy Strategy emphasises the important link between energy security and energy efficiency, a parallel synergy exists between regulatory, price and market reforms in Russia's gas sector and the effective implementation of any energy-efficiency policy. This is all the more evident if Russia is to realise benefits from its GHG emission reduction potential through investments in the framework of Kyoto Protocol flexibility mechanisms. To date, the Russian government has made positive steps in achieving the eligibility requirements. However, reinforced actions will be necessary in progressing its Action Plan in a timely way to build the confidence of the environmental community of Russia's resolve.

ANNEX 1. CALCULATION OF CONVERSION FACTORS

DIRECT METHANE EMISSIONS ALONG GAS TRANSMISSION AND DISTRIBUTION SYSTEMS

The factor that we use to convert 1 m³ of methane emissions into kilograms of CO_2e is calculated as follows:

16g CH_4/1mole × 1 mole/22.4 liter × 1 000 liter/m³ × 1 kg/1 000 g × 0.98 × 21 = 14.7 kgCO_2e/m³.

Where,

- 0.98 – is the assumed methane content of Russian natural gas. The range of methane content in natural gas in Russia is between 85% and 98% (*Energy Security of Russia,* 2005) or 93.5% (Hanle, 2003). We assume that gas in the transmission and distribution systems has been treated and purified and for this reason the high-end value of the range applies.[173]

- 21 – is the Global Warming Potential (GWP) of methane, indicated in the IPCC Guidelines. This is to be approved or modified by the new version of the IPCC Guidelines.

This conversion factor was rounded to 15 MtCO_2e/bcm which could lead to a slight overestimation of 2%. Given the range of uncertainties which exist in the emissions data, as well as those inherent in the values of methane content and GWP, this rounding error is not significant.

CO_2 EMISSIONS FROM NATURAL GAS COMBUSTION IN TRANSMISSION SYSTEMS

The factor used to convert 1 m³ of natural gas combusted into kilograms of CO_2 is calculated as follows:

15.3 tC/TJ × 34.4079 MJ/1000 m³ × 0.995 × 44/12 = **1.93 kgCO_2/m³**

---

173. The IPCC (2000) indicates 97.3% CH_4 content for typical gas analysis of North American gas transmission and distribution systems.

Where,

- 15.3 tC/TJ – the carbon emission factor of dry gas[174] indicated by IPCC (1997).

- 34.4079 TJ/1 000 m^3 – the net calorific value of Russian gas (IEA, 2005d).

- 0.995 – the fraction of carbon oxidised during combustion with 100% efficiency (IPCC, 1997). The lower combustion efficiency implies the direct release into the atmosphere of a part of the CH_4 contained in the natural gas. Given the high GWP of methane, the conversion factor could be significantly higher. A reduction of combustion efficiency of 5% (down to 95%) results in an increase in emissions of 30% (Hanle, 2003).

- 44/12 – the factor converting from tonnes of carbon to tonnes of CO_2.

This conversion factor was rounded to 2 MtCO$_2$/bcm (inducing the error of 3.5%). Given the conservative assumption of the combustion efficiency, this approximation is considered acceptable.

THE FLARING OF ASSOCIATED GAS

The calculation used to convert 1 m^3 of associated gas into kilograms of CO_2 equivalent is as follows:

For the combusted part of associated gas:

- Methane (CH_4, 75%): $Ef_{comb} \times 750 \text{ m}^3/1\ 000\text{m}^3 \times 44/22.4$

- Ethane (C_2H_6, 15%): $Ef_{comb} \times 2 \times 150 \text{ m}^3/1\ 000\text{m}^3 \times 44/22.4$

- Propane (C_3H_8, 4%): $Ef_{comb} \times 3 \times 40 \text{ m}^3/1\ 000\text{m}^3 \times 44/22.4$

- Butane (C_4H_{10}, 1%): $Ef_{comb} \times 4 \times 10 \text{ m}^3/1\ 000\text{m}^3 \times 44/22.4$

- CO_2 (3%): $30/22.4 \times 44$

For direct methane leaks to the atmosphere (*i.e.* due to incomplete combustion):

$(1\text{-}Ef_{comb}) \times C_{CH_4} \times (1\text{mole}/22.4\text{L}) \times (1\ 000\text{L}/1\text{m}^3) \times (16\text{g}CH_4/1\text{mole}) \times (1\text{kg}/1\ 000\text{g}) \times 21$

---

174. Dry gas is natural gas that does not require any hydrocarbon dew-point control to meet sales gas specification. However it may still require treating for water and acid gas content (IPCC, 2000).

Where,

- Ef_{comb} – the efficiency of combustion (see above). According to Hanle (2003), combustion efficiency is influenced by energy density of natural gas (MJ/m^3), which is a function of gas composition, and to a lesser extent, wind-speed. Hanle gives the approximate relationship between gas composition and combustion efficiency. We do not take into account the impact of wind speed in this study (no information available).[175]

- C_{CH_4} – the volumetric content of methane in associated gas in a "typical" Russian field. This value is uncertain given the potentially significant difference among fields and the age of the field. Other examples of the volumetric content of methane are available. For instance, Hanle (2003) uses the average for fields in the West Siberia Basin in the order of 74.2%.[176]

Given that the average efficiency of Russian flaring equipment can be rather high – about 95% efficiency, the conversion factor of 2.88 $kgCO_2e$ per m^3 is rounded to 2.9 $kgCO_2e/m^3$ with about a 1% error (see Figure 28).

Figure 28   Emission factors for associated gas flaring for different combustion efficiencies

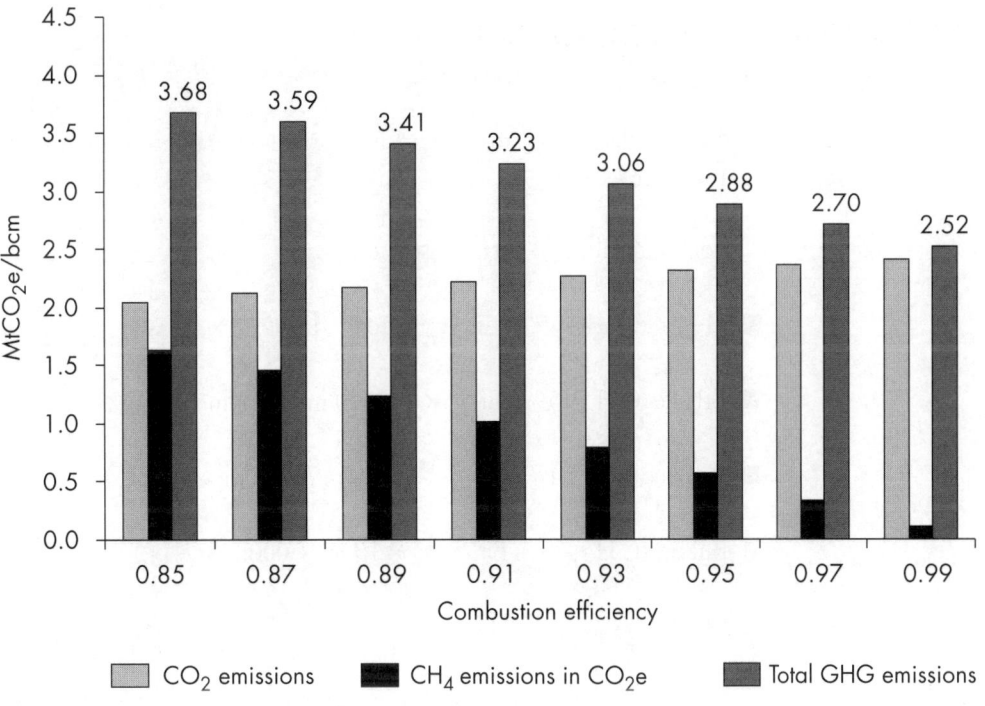

175. Hanle (2003) uses an 85% combustion efficiency.
176. The molecular content of carbon calculated based on measurements for the Rang Dong oil field (Vietnam) includes about 73% methane, 10% ethane and 2% of butane.

ANNEX 2. MAIN STAGES OF DEVELOPMENT OF RUSSIAN CLIMATE POLICY

Table 35   Historic timeline of Russian climate policy

	Significant landmarks
1994	Signature of the United Nations Framework Convention on Climate Change (UNFCCC) Establishment of the Inter-Agency Commission on Climate Change (ICCC) in charge of co-ordination and implementation (under Roshydromet's presidency)
1995	First National Communication to the UNFCCC Roshydromet is designated as the National Focal Point for the AIJ (Activities Implemented Jointly) Start of AIJ in Russia. From 1995 to 2002: 17 projects implemented, 9 registered by the UNFCCC Secretariat and 25 projects/plans not registered
1998	Second National Communication to the UNFCCC Federal Target Programme "Energy Savings in Russia for 1998-2005"
1999	Signature of the Kyoto Protocol to the UNFCCC
2000	Ministry of Economic Development and Trade (MEDT) is designated for co-presidency of the ICCC to re-inforce its political power
2001	Federal Target Programme "Energy-Efficient Economy for 2002-05 and until 2010" considering the Kyoto Protocol flexibility mechanisms as supplemental instruments for implementation of energy savings
2002	Third National Communication to the UNFCCC
2003	New Russian Energy Strategy until 2020 mentions the need to fulfil the Kyoto commitments of the Russian Federation
2004	Russian ratification of the Kyoto Protocol to the UNFCCC
2005	Comprehensive Action Plan to Implement the Kyoto Protocol in Russia including main developments of its national climate policy and the timing for its implementation
2005	Establishment of the Inter-Agency Commission on the Implementation of the Kyoto Protocol in Russia (MEDT is a leading ministry)

ANNEX 3. COMPRESSOR UNITS ALONG RUSSIA'S TRANSMISSION SYSTEM

Table 36 Structure of the stock of Russia's compressor units in January 2000

GPU type	Unit capacity, MW	Number of GPU	Total capacity, MW	Maximum service life, thou. h	Compressor units with maximum service life
Centaur	2.6	15	39	85	Elshano-Kurd (CSUGS*)
Centaur T-4500	3.0	18	54	–	Kasimovskaya
Centaur T-4500S	3.7	6	22	–	Uviazovskaya
GT-700-5	4.3	13	55	201	N. Tura
GTK-5	4.4	11	48	137	Berezanskaya
Taurus-60S	5.2	4	21	2	Severnaya
GT-750-6	6.0	37	222	164	Krasnoturyinskaya
Don-1	6.0	2	12	–	Istye
Avrora	6.0	50	300	–	Chaltyr
Ladoga	6.0	6	36	–	Voskresensk
GT-6-750	6.0	129	774	–	Long-Yugan
GTN-6	6.3	83	523	120	Vuktyl
GPA-Ts-6.3	6.3	409	2 577	145	Okhansk
Don 2	6.5	4	26	–	Ostrogozhsk
Don 3	6.5	1	7	–	Istye
GPA-Ts-6.3	8.0	25	200	–	Syzranskaya
GTNR-10	10.0	1	10	–	Algasovskaya
GTK-10	10.0	370	3 700	132	Algai
GTK-10-4	10.0	406	4 060	151	Pripoliarnaya
GTK-10	10.0	143	1 430	139	Frolovo
GPU-10	10.0	249	2 490	104	Sharan
GTK-10-1	10.0	7	70	–	Mikun
Coberra-182	11.9	18	214	118	Yarkovskaya
Coberra-182	12.9	11	142	102	Shatrovo
GPA-12R	12.0	12	144	20	Bardadymskoe
GTNR-12.5	12.5	1	13	42	Bardadymskoe
PGT-12C	16.0	5	80	–	Almaznaya
GTNR-16	16.0	1	16	6	Mokrous
GTN-16	16.0	58	928	71	Krasnoturyinskaya
GTN-16	16.0	2	32	22	Ukhta
GTN-16 -1	16.0	2	32	–	Privodinskaya
GPA-Ts-16	16.0	612	9 792	84	Torbeevskaya
GPA-Ts-16	18.0	24	432	–	Priozernaya
GPU-16	16.0	94	1 504	31	Komsomolskaya
GPA-16 Volga	16.0	3	48	–	Ordinskaya
GPA-16 G 90	16.0	20	320	109	Bogandinskaya
GPA-16 Zh 59	16.0	27	432	80	Mozhga
GTK-25I	25.0	61	1 525	132	Nadymskaya
GTK-25IR	25.0	40	1 000	83	Sechenovo
GTNR-25I	25.0	26	650	10	Mozhga
GPA-Ts-25	25.0	1	25	–	Tolyattinskaya
GTN-25	25.0	68	1 700	68	Zavolzhskaya
GTN-25-1	25.0	7	175	36	Donskaya
Total	–	**3 082**	**35 880**		
Electric-driven units	4.0 - 25.0	729	6 065		
Motor (diesel) drive compressors	0.7 - 5.5	212	266		
All drives		**4 023**	**42 211**		

*CSUGS – compressor station for underground gas storage
Source: Energy Security of Russia, 2005.

ANNEX 4. THE BEST-AVAILABLE TECHNOLOGY AND GAZPROM'S NEW COMPRESSOR UNITS

The direct comparison of efficiency parameters of the best-available technology (BAT) turbines and Gazprom's new facilities is complicated by possible differences in characterisation of the efficiency of units. In the case of Russian units, the efficiency may apply to the entire compressor unit including the drive engine. For the BAT, the efficiency applies only to the drive engine.

Table 37   The best-available compressor turbines with capacities from 4 to 31 MW

Manufacturer	Type	ISO-energy (MW)	Energy efficiency
BAT turbines with capacity of 4-12 MW:			
Solar	Centaur50	4.6	29.9%
GE Oil&Gas	GE5	5.6	31.5%
Solar	Taurus60	5.7	31.9%
Solar	Taurus70	7.7	34.8%
Siemens	Tornado	7.7	33.5%
Man Turbo	THM1304-10D	9.7	29.2%
Solar	Mars 90 S	9.9	33.1%
GE Oil&Gas	THM1304-11D	10.6	32.5%
Solar	Mars 100S	11.2	33.9%
Man Turbo	THM1304-11D	11.2	31.0%
GE Oil&Gas	GE10/2	11.7	32.6%
Average efficiency	–	–	*32.2%*
BAT turbines with capacity of 13-20 MW:			
Siemens	Cyclone	13.4	36.2%
GE Oil & Gas	PGT16	14.2	36.2%
Solar	Titan 130S	14.5	35.6%
GE Oil & Gas	PGT20	18.1	36.5%
Average efficiency	–	–	*36.1%*
BAT turbines with capacity of 21-30 MW:			
GE Oil & Gas	PGT25DLE	23.3	37.7%
Man Turbo	FT8-55 DLN	25.9	38.5%
Rolls Royce	RB211-6556	26.0	35.8%
Siemens	GT10C	30.1	37.3%
Average efficiency	–	–	*37.3%*

Source: Information as of July 2004.

Table 38 Gazprom's new generation of compressor units (compressors and drives)

Compressor unit type	Drive	Capacity, MW	Energy efficiency
Russian producers			
GPA-Ts-6.3	NK-14ST	6.3 (8)	30.0%
GTN-6U	GTN-6U	6.3 (8)	30.0%
GPA-12 Ural	PS-90GP1	12	34.0%
GPA-16 Ural	PS-90GP2	16	36.3%
GPA-Ts-16AL	AL-31ST	16	36.0%
GPA-Ts-16 Neva	AL-31ST	16	36.0%
PGT-21S	AL-31ST	16	36.0%
GPA-16 Volga	NK-38ST	16	36.5%
GTN-25-1	GTN-25-1	25	31.0%
GPA-Ts-25	NK-36ST	25	34.5%
GPA-25 Ural	PS-90GP25	25	39.4%
Baltika 25	GT-10	25	35.0%
Ukrainian producers			
GPA-Ts-6.3	DT-71	6.3	30.5%
GPA-Ts-6.3	DT-336	6.3	30.0%
GPA-16MG90	DG-90	16	34.0%
GPA-16MN80	DN-80	25	35.0%

Source: Gazprom, 2001b.

ANNEX 5. CH$_4$ EMISSION MEASUREMENT PROGRAMMES

GAS PRODUCTION AND PROCESSING

The Gazprom & Ruhrgas measurement programme examined three of the eight processing facilities of the Yamburggazdobitcha (Yamburg Gas Production Company) in northwestern Siberia.[177] Measurements took place at the oldest and newest processing plants (UKPG-2 built in 1986 and UKPG-4 built in 1994), and a gas condensate plant (UKPG-1 built in 1991). All relevant equipment, valves, pipelines, buildings and vents were checked for methane leaks.

The CH$_4$ emissions at the facilities and wellheads in Yamburg were relatively small. In 1996, the unintentional leaks of 35 Mm^3 of CH$_4$ amounted to only 0.02% of the gas produced by Yamburggazdobitcha. Operational methane leaks contributed to half of the measured CH$_4$ emissions.

Based on this study, CH$_4$ emissions from gas production and processing in Yamburg were estimated at about 0.06% of annual gas production.[178] The authors assume that the emission rates are representative of gas production in the whole region of West Siberia, which accounts for over two-thirds of Russian gas production (Dedikov *et al.*, 1999).

LINEAR PART OF PIPELINES

In 1996, Gazprom & Ruhrgas undertook a study using air-patrol methane leak detectors along 2 000 km of transmission pipelines of Tyumentransgaz. In 1997, the measurements were continued along another 630 km of transmission pipelines of Volgotransgaz at the Uzhgorod Corridor, including 350 valves, checked by foot patrol. Total pipeline emissions of Volgotransgaz amounted to 8 200 m^3/km/year (see Table 39). Emissions due to venting during repairs accounted for 58% of this total. Leaks represented about 33% of emissions and the remaining 9% was due to ruptures.

---

177. Facilities included in the study represented 30% of Yamburggazdobitcha output. In 2004, its production reached 241 bcm of gas or 45% of Gazprom's production.
178. Dedikov *et al.* (1999) include in the study only the direct methane release into the atmosphere which represent about 30% of operational methane emissions, the rest being flared according to operational data.

Table 39   Methane emissions from Volgotransgaz pipelines in 1997

Source of emissions along transmission pipelines	Emissions, m³/km/y	% of total
Leaks	2 700	33%
Repairs	4 800	58%
Ruptures	700	9%
Total	**8 200**	**100%**

Source: Dedikov et al., 1999.

In 2003, the Wuppertal Institute (2005) undertook measurements of emissions along 2 380 km of the export trunk pipeline system of the Central and Northern Corridors which cover a total of 3 376 km and 3 075 km, respectively, linking the production regions of West Siberia to Germany and Western Europe (see table 40).[179] The pipelines were surveyed from the air by helicopter. The measurements covered also 25 intersection valves installed on the gas pipeline that belong to the compressor stations at intervals of approximately 15-30 km.

The Wuppertal Institute estimated 6 458 m³/km per year of CH_4 emissions along the pipelines examined. This is 20% less than the earlier estimates made by the Gazprom & Ruhrgas study. The structure of emissions reflected in the two studies is quite similar. Gas vented before maintenance and repair accounted for over 58% of emissions, leaks represented another 38%, while ruptures contributed only 4% of total.[180] The Wuppertal Institute study attributes the significantly reduced emissions due to accidents (a halving of emissions compared to the 1996 Gazprom & Ruhrgas results) to improvements in diagnosis and preventive repairs by Gazprom in the interim period.

Table 40   Export gas pipelines surveyed by the Wuppertal Institute in 2003

Regional branch	Compressor stations	Built	Length surveyed	Intersection valves	Emissions
Mostransgaz	Davidovskaya	1983-1988	300 km	1	
	Kursk	1983-1988	300 km	4	
Severgazprom	Uchta	1969-1977	1 200 km	6	
	Njukzeniza	1969-1981	580 km	8	
Tyumentransgaz	–	–	–	6	
Total	**–**	**–**	**2 380**	**25**	**6 458 m³/km/y**

Source: Wuppertal Institute, 2005.

---

179. These consist mainly of 4-6 parallel pipelines. The Central Corridor is 22 000 km long while the Northern Corridor is 12 000 km.
180. This emission factor is calculated according to the accident statistics provided by Gazprom.

ANNEX 6. STUDIES ON METHANE EMISSIONS AT COMPRESSOR STATIONS

Table 41   Estimates of methane emissions at Russian compressor stations

Reference/ source	Place / Date of commissioning	Drives	Methane emissions	Measurements
US EPA & Gazprom (1996)	Chaplygin	8 x 6.3MW GT 2x16MW GT		Field & valve yard & blowdown unit/valve vents. Not compressors themselves
	Pervomaiskaya	3x25MW GT 28x19.5MW EC		
	Petrovsk	6x6MW GT 25x4MW EC		
	Storojovka	5x6.3MW GT 7x4MW EC 4x6.3MW EC		
Gazprom & Ruhrgas (Dedikov *et al.*, 1999)	Kazym 1971-77	222 MW	75 000 m³/MW/y	Operational vents and unintentional leaks measured in 1996 and 1997
	Upper Kazym 1983-97	40 units with total 715 MW	53 000 m³/MW/y	
Gazprom & TransCanada (Venugopal *et al.*, 2003)	Pochinki, Torbeevo, 2001*	6x25MW 13x15MW 10x19MW 3x14x11 MW	21 364 m³/MW/y	Measurement of valve leakage in 2001
Wuppertal Institute (2005)	Davidovskaya, 1985	7x12.5MW EC	49 418 m³/MW/y	Operation-related emissions (5 227 m³/MW/y) and unintentional leaks (44 191 m³/MW/y)
	Kurskaya, 1985	3x22.2MW GT		
	Uchta, 1982, 2001	6x10MW GT 2x16MW GT		
	Njukzeniza, 1986, 1987-88, 2001	5x6MW GT 13x10MW GT 2x16MW GT		
	Kazym 1972, 1977	6x6MW GT 6x10MW GT		

GT - gas turbines, EC - electric compressors
* Total capacity 1 008 MW EC&GT

REFERENCES

Action Plan (2005): Comprehensive Action Plan to Implement the Kyoto Protocol in the Russian Federation (Kompleksniy plan deystviy po realizatsi Kiotskogo Protokola v Rossiyskoy Federatsii), MEDT (Ministry of Economic Development and Trade of the Russian Federation) Web site.

Altfeld, K. *et al.* (2000), "Methane Emissions Caused by the Gas Industry World-Wide", Report of Study Group 8.1 presented at the 21st World Gas Conference, Nice, France, 6-9 June.

Aslanyan, G., O. Pluzhnikov (2003), "Russian View on Kyoto Protocol Perspectives" (Perspektivi Kiotskogo protokola - vzgliad iz Rossii), presented at the WNA meeting, Moscow, 13 May.

Astakhov, V. (2005), "Options for Reducing the Volumes of Associated Gas Flaring during the Production at the Lukoil Oil Fields" (Puti snizhenia ob'emov szhigania neftianogo gaza pri razrabotke neftianih mestorozhdeniy OAO "Lukoil"), presented at the EU-Russia Technology Centre Round Table "Modern Technologies and Practices for Decreasing the Flaring of Associated Gas", Moscow, 19 April.

Babiker, M.H. *et al.* (2002), "The evolution of a Climate Regime: Kyoto to Marrakesh", *MIT Report*, No. 82.

Baron, R. (1999), "Market Power and Market Access in International GHG Emissions Trading", *OECD/IEA Information Paper*, Paris, www.oecd.org/dataoecd/16/61/2391156.pdf.

BASREC (Baltic Sea Region Energy Co-operation) (2003), *Co-operation in the Energy Sector of the Baltic Region Countries: Guidelines for the Joint Implementation*, www.basrec.org.

Blanchard, O., P. Criqui, A. Kitous (2002), "Après La Haye, Bonn et Marrakech: le futur marché international des permis de droits d'émissions et la question de l'air chaud", *Cahier de Recherche* de l'IEPE, No. 27, http://web.upmf-grenoble.fr/iepe/textes/Cahier27.pdf.

Blyth, W., R. Baron (2003), "Green Investment Schemes: Options and Issues", OECD/IEA, Paris, www.oecd.org/dataoecd/48/54/19842798.pdf.

Blyth, W., M. Bosi (2004), "Linking non-EU Domestic Emissions Trading Schemes with the EU Emissions Trading Scheme", IEA/OECD, Paris, www.oecd.org/dataoecd/38/7/32181382.pdf.

Budzulyak (2004), "Reconstruction – a Way to Increase Capacity of the Russian Gas Transmission System" (Rekonstruktsia – put povishenia proizvodstvennoy mochnosti gazotransportnoy sistemi Rossii), presentation of Gazprom at the 2nd International Forum "Gas of Russia-2004".

Budzulyak B., Ch. P. Bechervordersandfort (2004), "Measuring of Methane Emissions in Russia", presented at the Eurogas Meeting, Brussels, 30 March.

C^3 VIEW (2005), "BP Canada Improves Compressor Performance", C^3 VIEW, Issue No. 13, January.

CAPP (Canadian Association of Petroleum Producers) (2003), *Calculating Greenhouse Gas Emissions*,
www.capp.ca/raw.asp?NOSTAT=YES&dt=PDF&dn=55904.

Cedigaz (2004), *Natural Gas in the World*, Institut Français du Pétrole, Rueil-Malmaison.

CENEf (Centre for Energy Efficiency), PNNL (Pacific Northwest National Laboratory) (2004), *National Inventory of Energy Related Emissions of Greenhouse Gases in Russia*, Moscow.

CEPA (Canadian Energy Pipeline Association) (2003), "Ninth Report to Canada's Climate Change Voluntary Challenge & Registry Inc.",
www.cepa.com.

CDM EB (Clean Development Mechanism Executive Board) (2004), *Tool for the Demonstration and Assessment of Additionality*, UNFCCC,
http://cdm.unfccc.int/methodologies/PAmethodologies/AdditionalityTools/Additionality_tool.pdf.

CDM EB (2005a), "Clarifications on the Consideration of National and/or Sectoral Policies and Circumstances in Baseline Scenarios", Annex 3, *Meeting Report* No. 22,
http://cdm.unfccc.int/EB/Meetings/022/eb22_repan3.pdf.

CDM EB (2005b), *Leak Reduction from Natural Gas Pipeline Compressor or Gate Stations*, Approved Methodology AM00023,
http://cdm.unfccc.int/UserManagement/FileStorage/CDMWF_AM_285525461.

CDM EB (2005c), *Recovery and Utilisation of Gas from Oil Wells that would otherwise be Flared*, Revision to Approved Baseline Methodology AM0009,
http://cdm.unfccc.int/UserManagement/FileStorage/AM0009version2.pdf

Chang, S. (2001), "Comparing Exploitation and Transportation Technologies for Monetisation of Offshore Stranded Gas", paper presented at the SPE (Society of Petroleum Engineers) Asia Pacific Oil and Gas Conference, Jakarta, 17-19 April.

COP/MOP 1 (Conference of the Parties to the UNFCCC serving as the Meeting of the Parties to the Kyoto Protocol) (2005), *Guidelines for the implementation of Article 6 of the Kyoto Protocol*, Decision 9/CMP.1, Appendix B, Criteria for baseline setting and monitoring,
http://unfccc.int/resource/docs/2005/cmp1/eng/08a02.pdf#page=2.

Dedikov, E.V. *et al.* (1999), "Estimating Methane Releases from Natural Gas Production and Transmission in Russia", *Atmospheric Environment* No. 33, pp. 3291-3299.

Dedikov, E.V. *et al.* (2000), "The AIJ Project of Ruhrgas AG and OAO Gazprom and its Evaluation", Gazprom paper presented at the 21st World Gas Conference, Nice, France, 6-9 June.

Delbeke J. (2005), "The Final Implementation Challenge", *Carbon Market Europe,* Point Carbon, 21 October.

Enbridge Gas Distribution Inc. (2003), "Greenhouse Gas Emission Action Plan: Update 2002",
www.ghgregistries.ca/registry/out/C0052-ENBRIDGEGD-03-PDF.PDF.

Enbridge Gas Distribution Inc. (2004), "Greenhouse Gas Emissions Inventory and Management Plan - 2004",
www.ghgregistries.ca/registry/out/C0052-EGD_AP04-PDF.PDF

Energy-Saving Economy (2001): Federal Target Programme of the Russian Federation "Energy-Saving Economy for 2002-05 and through 2010", Resolution of the Government of the Russian Federation No. 83-r of 22 January 2001 (Rasporiazhenie Pravitelstva RF ot 22.01.2001 "Federalnaya tselevaya programma "Energoeffektivnaya ekonomika na 2002-05 i na perspektivu do 2010").

Energy Security of Russia (2005): *Security of Russia, Energy Security: Gas industry of Russia* (Bezopasnost Rossii, Energeticheskaya bezopasnost: Gazovaya promishlennost Rossii), Moscow.

Energy Strategy (2003): "Energy Strategy of Russia for the Period through 2020", Resolution of the Government of the Russian Federation No. 1234-r of 28 August (Rasporiazhenie Pravitelstva RF ot 28.08.2003 "Energeticheskaya Strategia Rossii na period do 2020"),
www.minprom.gov.ru/docs/strateg/1.

Federal Mineral Resource Act (1992): *The Federal Mineral Resource Act of the Russian Federation No. 2395-I of 21 February* (Zakon Rossiyskoy Federatsii "O Nedrah" ot 21.02.1992 No. 2395-I),
www.garant.ru/law/10004313-000.htm.

Fernandez, R., D. Lieberman, D. Robinson (2004), "US Natural Gas STAR Programme: Success Points to Global Opportunities to Cut Methane Emissions Cost-Effectively", *Oil & Gas Journal,* 12 July,
www.epa.gov/gasstar/news/interops.htm.

FTS of Russia (Federal Tariff Service of the Russian Federation) (2005), Information Letter No. SN-3923/9 of 28 June (Informatsionnoe pismo FST Rossii ot 28.06.2005 No. SN-3923/9), www.fstrf.ru/press/info/1.

Gale, J., P. Freund (2000), "Reducing Methane Emissions to Combat Global Climate Change: The Role Russia Can Play", presented at the Second International Methane Mitigation Conference, Novosibirsk, Russia, 18-23 June.

Gavrilov, V.V. (2005a), "Kyoto transactions are already possible, but only as options" (Zakluchat sdelki v ramkah Kiotskogo Protokola mozhno uzhe seychas, no tolko v vide optsionov), *Interfax,* 19 August, MEDT Web site.

Gavrilov, V.V. (2005b), "Would be Unfortunate not to Achieve an Agreement" (Budet obidno, esli mi ne dogovorimsya), *Itogi* No. 33, 15 August, MEDT Web site.

Gazprom (1998), *Strategic Development of the Russian Gas Industry,* Gazoil Press, Moscow.

Gazprom (2001a), *Materials of the Bureau of the Scientific and Technical Board of Gazprom on its Energy-Saving Programme for the period 2001-2010,* (Materiali Bureau Nauchno-Teknicheskogo Soveta OAO "Gazprom" o Kontseptsii Energosberezhenia v OAO "Gazprom" na 2001-2010), Moscow.

Gazprom (2001b), *Gazprom Energy-Saving Programme for the Period 2001-2010* (Kontseptsia Energosberezhenia v OAO "Gazprom" na 2001-2010), Moscow.

Gazprom (2003), *Annual Report 2003,* www.gazprom.com/documents/gazprom_eng.pdf.

Gazprom (2004a), *Annual Report 2004,* www.gazprom.com/documents/Annual_Report_Eng_2004.pdf.

Gazprom (2004b), *Environmental Report 2004* (Ekologicheskiy otchet OAO "Gaprom" za 2004), www.gazprom.ru/documents/Ecology.pdf

Gazprom (2004c), "Comprehensive Programme of Reconstruction and Technical Modernisation of the Gas Distribution Network is Approved by the Board" (Pravlenie odobrilo Komple-ksnuyu programmu rekonstruktsii i teknicheskogo perevooruzhenia gazovogo hoziastva), Communication, 30 March, www.gazprom.ru/news/2004/03/302144_11129.shtml.

Gazprom (2005a), *Gazprom in Numbers 2000-2004* (Gazprom v tsifrah 2000-2004), www.gazprom.ru/documents/Statistika%20Rus.pdf.

Gazprom (2005b), *Gazprom in Questions and Answers* (Gazprom v voprosah i otvetah), www.gazpromquestions.ru.

Gazprom, US EPA (United States Environment Protection Agency) (1996), *Methane Leak Measurements at Selected Natural Gas Pipeline Compressor Stations in Russia* (Draft),
www.epa.gov/gasstar/pdf/Scan003_508.pdf.

GGFR (Global Gas Flaring Reduction Public-Private Partnership) (2003), *Kyoto Mechanisms for Flaring Reductions,* Report No. 2, World Bank Group.

GGFR (2004a), *Regulation of Associated Gas Flaring and Venting; A Global Overview and Lessons from International Experience,* Report No. 3, World Bank Group.

GGFR (2004b), *A Voluntary Standard for Global Gas Flaring and Venting Reduction,* Report No. 4, World Bank Group.

GGFR (2004c), *Flared Gas Utilisation Strategy, Opportunities for Small-Scale Uses of Gas,* Report No. 5, World Bank Group.

GGFR (2005a), *Expanded Update November 2005,* Summary for the November Steering Committee Meeting.

GGFR (2005b), *Framework for Clean Development Mechanism (CDM) Baseline Methodologies,* Report No. 6, World Bank Group.

GoF (Global Opportunities Fund) (2005), "Examples of Successful Projects: Support for the Russian Gas Industry to Participate in Kyoto Mechanisms",
www.fco.gov.uk/servlet/Front?pagename=OpenMarket/Xcelerate/ShowPage&c= Page&cid=1122472523912.

Golub, A. *et al.* (2004), "Breaking through Barriers in Russia", *Environmental Finance,* May.

Golub A., D. Mercellino (2005), "What is to be Done?" *Carbon Finance,* December 2004/January 2005.

Gosgorteknadzor (2003), *Approval of Security Rules for the Gas Distribution and Gas Consumption Systems,* Decision No. 9 of 18 March 2003 (Postanovlenie Gorgorteknadgora RF ot 18.03.2003 No. 9 "Ob utverzhdenii pravil bezopasnosti sistem gazorasperedelenia i gazopotreblenia").

Gosgorteknadzor (2005): Federal Mining and Industrial Inspection (2005), *Information Bulletin of Gosgorteknadgor of Russia* (Infomatsionniy Bulleten Gosgorteknadzora Rossii, Federalnaya Sluzhba po Ekologicheskomu, Teknologicheskomu i Atomnomu Nadzoru), No. 2(17).

Haites, E. (2004), "Estimating the Market Potential for the Clean Development Mechanism: Review of Models and Lessons Learned", PCF *plus* report No. 19, IEA/ IETA/PCF, Washington DC, June 2004,
www.iea.org/textbase/publications/free_new_Desc.asp?PUBS_ID=1027.

Hanle, L. (2003), "Emission Reductions in the Natural Gas Sector through Project-Based Mechanisms", *IEA Information Paper,* www.iea.org/texbase/papers/2003/devbase.pdf.

Hasselknippe, H. (2005), "Development in the Carbon Market", presented at Chatham House Conference "Emerging Carbon Markets", 16 June.

IBRC (International Business Relations Corporation) (2003), "Analysis of Natural Gas Pipeline Compressor Equipment Market in Russia", Moscow.

ICCC (Inter-Agency Commission of the Russian Federation on Climate Change) (2002), *Third National Communication of the Russian Federation,* Moscow, http://unfccc.int/resource/docs/natc/rusnce3.pdf.

IEA (2001), *International Emissions Trading: From Concept to Reality,* OECD/IEA, Paris.

IEA (2002), *Russian Energy Survey 2002,* OECD/IEA, Paris.

IEA (2003), *World Energy Investment Outlook: 2003 Insights,* OECD/IEA, Paris.

IEA (2004a), *Coming in from the Cold, Improving District Heating Policy in Transition Economies,* OECD/IEA, Paris.

IEA (2004b), *World Energy Outlook 2004,* OECD/IEA, Paris.

IEA (2004c), *Energy Efficiency in Economics in Transition (EITs): A Policy Priority,* OECD/IEA, Paris.

IEA (2005a), *Act Locally, Trade Globally,* OECD/IEA, Paris.

IEA (2005b), *CO_2 Emissions from Fuel Combustion 1971-2003,* OECD/IEA, Paris.

IEA (2005c), *Energy Statistics of non-OECD Countries,* OECD/IEA, Paris.

IEA (2005d), *Natural Gas Information,* OECD/IEA, Paris.

IETA (International Emissions Trading Association) (2006), *IETA's Guidance note through the CDM Project Approval Process,* Version 2.0, May, www.ieta.org/ieta/www/pages/getfile.php?docID=900.

IPCC (Intergovernmental Panel on Climate Change) (1997), *Revised 1996 IPCC Guidelines for National Greenhouse Gas Inventories,* National Greenhouse Gas Inventory Programme, www.ipcc-nggip.iges.or.jp/public/gl/invs1.htm.

IPCC (2000), *Good Practice Guidance and Uncertainty Management in National Greenhouse Gas Inventories,* National Greenhouse Gas Inventory Programme, www.ipcc-nggip.iges.or.jp/public/gp/english.

IPIECA (International Petroleum Industry Environmental Conservation Association), OGP (International Association of Oil and Gas Producers), API (American Petroleum Institute) (2003), *Petroleum Industry Guidelines for Reporting Greenhouse Gas Emissions,*

www.ogp.org.uk/pubs/349.pdf.

Karasevich A., A. Terekhov (2004), "Energy Savings in Natural Gas Distribution and Utilisation in the Russian Federation", presentation by Promgaz and Regiongazholding, 16 March,

http://unece.org/ie/se/pdfs/GasS&U_Karacevich.pdf.

Kexel, D.T. (2005), "Projects in Gas Sector: Methodology and Data Requirements", presented at the Seminar on "Preparation of Carbon Finance Projects: Project Cycle, Documentation, Methodologies", World Bank Carbon Finance, Moscow, 9-10 August,

http://194.84.38.65/mdb/cmsitems/375/DK_Projects_Gas_Sector_eng.pdf.

Khanti-Mansiysk Administration (2005), "Analysis of the Utilisation of the Associated Petroleum Gas in the Khanti-Mansiysk Autonomous Region – Yugra in 2004" (Analiz ispolizovania poputnogo neftyanogo gaza po Khanti-Mansiyskomu Avtonomnomu okrugu – Yugre v 2004), presented at the EU-Russia Technology Centre Round Table "Modern Technologies and Practices for Decreasing the Flaring of Associated Gas", Moscow, 19 April.

Kirillov, D. (2005), "Gas Transportation Rules" (Pravila gazovogo dvizhenia), *Gazprom Journal* No. 1-2, pp. 22-24.

De Klerk, L. (2003), "The Dutch ERUPT Programme – Materialising the Russian Potential", presented at the Climate Strategies Workshop "Implementing Kyoto in Russia and the CIS: Moving from theory to practice", Moscow, 8-9 April,

www.climate-strategies.org.

Kokorin, A.O., I.G. Gritsevitch, G.V. Safonov (2004), *Climate Change and Kyoto Protocol: Realities and Opportunities* (Izmenenie klimata i Kiotskiy Protokol: realii i prakticheskie vozmozhnosti), WWF, Moscow.

Kokorin, A.O. (2005), "Methodologies for Methane Emissions Inventories" (Metodologii inventarizatsii utechek metana), presented at the Conference on the Dissemination of Methane Emission Control Experience in the Gas Industry, Higher School of Economics, Moscow, 11 March.

Korpoo A. (2005), "Russian Energy-Efficiency Projects: Lessons Learned from Activities Implemented Jointly Pilot Phase", *Energy Policy* No. 33, pp. 113-126.

Kuraev, S.N., G.V. Safonov (2005), "Assistance to the Gas Industry in its Participation in the Kyoto Protocol" (Sodeystvie gazovoy otrasli v uchastii v Kiotskom Protokole), presented at the Conference on the Dissemination of Methane Emission Control Experience in the Gas Industry, Higher School of Economics, Moscow, 11 March.

Kursk Methane Emission Reduction Options (2005) (Proekti sokrachenia emissii metana po Kurskoy oblasti), Kursk.

Kursk JI PDD (2005): Project Design Document for a Joint Implementation Project "Methane Emissions Avoidance in the Kursk Gas Distribution Network", www.dnv.com/certification/climatechange/Upload/PDD%20Kursk%20Project-2005-12-30.pdf.

Kurskgaz (2005), *Trimestrial Report of Kurskgaz: 2nd Trimester of 2005* (Ezhekvartalniy otchet OAO "Kurskgaz" za 2 kvartal 2005), Kusrk, www.kurskgaz.ru.

Laouri, F., E. Tellegen, K. Tourilova (2004), "Joint Implementation in Energy between the EU and Russia: Outlook and Potential", *Energy Policy,* No. 32, pp. 899-914.

Leneva, M.E. (2002), *Possible Approaches and the Next Steps for the Development of a National Inventory System in the Russian Federation,* Centre for Environmental Economic Research and Information, Moscow, www.oecd.org/dataoecd/5/37/2467501.pdf.

Leontiev, E.V., O.P. Stureiko (2003), "Role of Inspection in the Development and Implementation of Reconstruction Programmes for the Gas Transmission System", presented at the Conference "Diagnostics 2003", Malta, 10 April.

Leontiev, E.V. *et al.* (2003a), "OAO Gazprom's Participation in an International Ecological Project on Greenhouse Gas Emission Reduction", presented at the World Gas Conference, Tokyo, 1-5 June.

Leontiev, E.V., O.P. Stureiko, V.A. Schurovsky (2003b), "Strategy of the Reconstruction of the Gas Transportation System of Gazprom" (Strategia rekonstruktsii gazotransportnoy sistemi OAO Gazprom), *Gazovaya Promishlennost,* No. 10, pp. 63-66.

Marrakech Accords (2001): UNFCCC (United Nations Framework Convention on Climate Change) (2001), *The Marrakesh Accords, Report of the Conference of the Parties on its Seventh Session,* Addendum, Part Two, Vol. 2, Marrakesh, Morocco, 29 October – 10 November, http://unfccc.int/resource/docs/cop7/13a02.pdf

Maslennikov, D. (2004) "Gas Transportation: Satisfactory State" (Transportirovka: Pusl v norme), *Gazprom Journal,* No. 1, pp. 12-13.

MEDT (Ministry of Economic Development and Trade of the Russian Federation) (2006a), "Implementation of the Kyoto Protocol to the UNFCCC by the Russian Federation" (O hode vipolneniya Kiotskogo Protocola k Ramochnoy Konventsii OON ob izmenenii klimata), Materials for the Session of the Russian Government on 16 March 2006, MEDT Web Site.

MEDT (2006b), "Draft Programme of Social and Economic Development of Russia for the Medium Term from 2006 to 2008" (Proekt Programmi Sotsialnogo i Ekonomicheskogo Razvitia Rossii na Srednesrochnuyu Perspektivu, 2006-2008), MEDT Web site.

Mielke E. *et al.* (2004), *Russia and Kyoto – Match Made in Heaven?* Dresdner Kleinwort Wasserstein Research, London.

Müller, B. (2004), "The Kyoto Protocol: Russian Opportunities", *Briefing note RIIA,* Oxford Institute for Energy Studies, March,

www.wolfson.ox.ac.uk/~mueller/.

NATSOURCE (2003), "Governments as Participants in International Markets for Greenhouse Gas Commodities", EPRI/IEA/IETA/IDDRI,

www.iddri.org/iddri/telecharge/climat/rapport_natsource.pdf.

Natural Gas STAR (2003a), *Directed Inspection and Maintenance at Compressor Stations,* www.epa.gov/gasstar/pdf/lessons/ll_dimcompstat.pdf.

Natural Gas STAR (2003b), *Replacing Wet Seals with Dry Seals in Centrifugal Compressors,* www.epa.gov/gasstar/pdf/lessons/ll_wetseals.pdf

Natural Gas STAR (2003c), *Directed Inspection and Maintenance at Gate Stations and Surface Facilities,*

www.epa.gov/gasstar/pdf/lessons/ll_dimgatestat.pdf.

Natural Gas STAR (2004a), *Reducing Emissions when Taking Compressors Off-line,* www.epa.gov/gasstar/pdf/lessons/ll_compressorsoffline.pdf

Natural Gas STAR (2004b), *Minimising Emissions from Pipeline Blowdowns,* Natural Gas Star Partner Update, Summer,

www.epa.gov/gasstar/pdf/6-04newsletter.pdf.

Odisharia, G.E., V.S. Safonov, V.I. Ezunenko (2003), "Security Indicators and Risk Analysis in the Exploitation of Gazprom Gas Transportation Systems" (Pokazateli bezopasnosti i analiz riska pri ekspluatatsii ob'ektov gazotransportnih system OAO "Gazprom"), presented at the World Gas Conference, Tokyo, 1-5 June.

OECD (2005), *OECD Reviews of Regulatory Reform: Russia,* Paris.

Payusov Yu. V. (2005), "Utilisation of Associated Gas at Yukos: Problems and Solutions", presented at the EU-Russia Technology Centre Round Table "Modern Technologies and Practices for Decreasing the Flaring of Associated Gas", Moscow, 19 April.

PCF (Prototype Carbon Fund) (2000), "Baseline Methodologies for PCF Projects", *PCF Implementation Note* No. 3,

http://carbonfinance.org/docs/pcfin_3_baselines_6dec00.doc

PCF (2002), "Policy Framework for Obtaining Early Credit for Emission Reductions in JI Projects", *Draft PCF Implementation Note* No. 8,

http://carbonfinance.org/docs/ImplementationNote8.doc.

Platonova, A. (2005), "Investment Projects in the Russian Oil and Gas Industry in the Framework of the Kyoto Protocol Flexibility Mechanisms" (Projets de l'investissement dans l'industrie pétrolière et gazière russe dans le cadre des mécanismes de flexibilité du Protocole de Kyoto), PhD Thesis, ENSPM/IFP, Université de Bourgogne, Rueil-Malmaison.

Plotnikov, V.S. (2005), "On Problems of Associated gas Utilisation in the Tyumen Region" (O problemah ispolzovania neftyanogo poputnogo gaza (NPG) v Tyumenskoy Oblasti), OAO "Gazprom", OAO "Sibur", presented at the EU-Russia Technology Centre Round Table "Modern Technologies and Practices for Decreasing the Flaring of Associated Gas", Moscow, 19 April.

PNNL (Pacific Northwest National Laboratory) (2004), *Kazakhstan GHG Emissions Inventory from Gas Transportation and Distribution, 1990 2002,* Final Project Report.

Point Carbon (2004), "Ruhrgas Aims to Boost CO_2 Credits with Russian JI Projects", *Point Carbon,* 19 March.

Popov, I. (2001), "Estimating Methane Emissions from the Russian Natural Gas Sector", Advanced International Studies Unit, PNNL,

www.pnl.gov/aisu/pubs/Gazprom.PDF.

Pravosudov, S. (2004b), "Unifying Distribution Networks" (Ob'edinayem seti), *Gazprom Journal,* No. 12, pp. 6-9.

Pravosudov, S. (2004a), "Investment Openness" (Investitsionnaya otkritost), *Gazprom Journal,* No. 2, pp. 6-9.

Remizov, V.V. *et al.* (2000), "Strategy of Methane Emission Reduction in the Gas Industry of Russia", paper presented at the Second International Conference on Methane Mitigation, Novosibirsk, 18-23 June.

Renaissance Capital (2002): Landes A., V. Mentev (2002), *Gazprom: The Dawning of a New Valuation Era: Re-Initiation of Coverage,* Renaissance Capital.

RIATEC (2006), "Gazprom postpones the industrial development of Yamal by 10-12 years ("Gazprom" otkladivaet promishlennoe osvoenie Yamala na 10-12 let), 10 April.

Robinson, D.R., R. Fernandez, R.K. Kantananeni (2002), "Methane Emissions Mitigation Options in the Global Oil and Natural Gas Industries", www.epa.gov/gasstar/pdf/ng020.pdf.

RREC (Russian Regional Environmental Centre) (2003), *Independent Evaluation of the Consequences of the Ratification of the Kyoto Protocol by Russia* (Nezavisimaia otsenka posledstviy prisoednenia Rossii k Kiotskomu protokolu), Moscow.

Ruhrgas, Gazprom (1997), "Modelling and Optimisation of Grid Operation of the Gas Transportation System "Ushgorod Corridor" of Wolgotransgas (Gazprom)", Activities Implemented Jointly under the Pilot Phase, http://unfccc.int/kyoto_mechanisms/aij/activities_implemented_jointly/items/1732.php.

Russian Government (2001a): The Ordinance of the Government of the Russian Federation No. 334 of 3 May 2001 "On ensuring access to independent organisations to the gas transportation system of Gazprom" (Postanovlenie Pravitelstva RF ot 3.05.2001 No. 334 "Ob obespechenii nediskriminatsionnogo dostupa k gazotransportnim sistemam").

Russian Government (2001b): The Ordinance of the Government of the Russian Federation No. 921 of 29 December 2001 "On approving the rules for establishing normative losses in natural resource development" (Postanovlenie Pravitelstva RF ot 29.12.2001 No. 921 "Ob utverzhdenii pravil utverzhdenia normativov poter poleznih iskopaemih pri dobiche, teknologicheski sviazannih s priniatoy shemoy i teknologiey razrabotki mestorozhdenia").

Russian Government (2005): The Ordinance of the Government of the Russian Federation No. 410 of 1 July 2005 "On the modification of Annex No. 1 to the Government Decree No. 344 of 12 July 2003 on the fees for pollutant emissions and discharges (…)" (Postanovlenie Pravitelstva RF ot 01.07.2005 No. 410 "O vnesenii izmeneniy v prilozhenie No. 1 k Postanovleniu Pravitelstva ot 13.06.2003 No. 344).

Russian Government (2006a), The Resolution of the Government of the Russian Federation No. 215-r of 2 February 2006 "On the Russian GHG emission Registry" (Rasporiazhenie Pravitelstva RF ot 20.02.2006 No. 215-r o sozdanii rossiyskogo reestra uglerodnih edinits).

Russian Government (2006b), The Resolution of the Government of the Russian Federation No. 278-r of 1 March 2006 "On the Russian National System for the Estimation of GHG Emissions" (Rasporiazhenie Pravitelstva RF ot 01.03.2006 No. 278-r "O sozdanii rossiyskoy sistemi otsenki vibrosov parnikovih gazov").

Safonov, G.V. (2004), *Ideas about Fugitive Emission Reductions in Russia,* Note, Moscow.

Seleznev, K. (2004), "Networks become obsolete, but it's Gazprom who has to worry" (Seti prihodyat v negodnost, a golova bolit u "Gazproma"), *Vedomosti,* 5 August, www.gazprom.ru/comments/2004/08/051413_13428.shtml.

Silva L.C. *et al.* (2004), "Report on the In-depth Review of the Third National Communication of the Russian Federation", FCCC/IDR.3/RUS, UNFCCC, Geneva, http://unfccc.int/national_reports/annex_i_natcom/idr_reports/items/2711.php.

Streck Ch. (2005a), "Too Many Mechanisms, Too Few Institutions: Challenges and Chances for EITs", presented at the SBSTA (Subsidiary Body for Scientific and Technological Advice) Meeting, Bonn, 25 May.

Streck Ch. (2005b), "Joint Implementation: History, Requirements, and Challenges", in ed. Freestone D., Ch. Streck (eds.), *Legal Aspects of Implementing the Kyoto Mechanisms,* Oxford University Press.

Tangen, K. *et al.* (2002), *A Russian Green Investment Scheme, Securing Environmental Benefits from International Emissions Trading,* Climate Strategies, www.climate-strategies.org/gisfinalreport.pdf.

TransCanada (2002), *8th Report to the Voluntary Challenge & Registry: 2002 Submission,* www.ghgregistries.ca/registry/out/C0135-TRANSCAN-02-PDF.PDF.

TransCanada (2003), *2003 Report to Canada's Climate Change Voluntary Challenge & Registry,* www.ghgregistries.ca/registry/out/C0135-TC-03V2-PDF.PDF.

TROIKA DIALOG (2006), *Russia Market Daily,* 9 March.

US EIA (United States Energy Information Administration) (1997), *Oil and Gas Resources of the West Siberian Basin, Russia, Appendix D: Field Summaries,* www.eia.doe.gov/pub/oil_gas/natural_gas/analysis_publications/oil_gas_resources_siberian_basin/pdf/apdx_d.pdf.

US EPA (2001), *Non-CO$_2$ Greenhouse Gas Emissions from Developed Countries: 1990-2010,* EPA-430-R-01-007, www.epa.gov/methane/pdfs/fulldocumentofdeveloped.pdf.

US EPA, GRI (Gas Research Institute) (1996a), *Methane Emissions from the Natural Gas Industry: Executive Summary,* Vol. 1, Washington, DC, Natural Gas STAR Web site.

US EPA, GRI (1996b), *Methane Emissions from the Natural Gas Industry: Underground Pipelines,* Vol. 9, Washington, DC, Natural Gas STAR Web site.

US EPA, GRI (1996b), *Methane Emissions from the Natural Gas Industry: Metering and Pressure Regulating Stations in Natural Gas Transmission and Distribution,* Vol. 10, Washington, DC, Natural Gas STAR Web site.

US NOAA (United States National Oceanic and Atmospheric Administration) (2005), "Composite of clear night images of West Siberia from the Defense Meteorological Satellite Programme (DMSP) satellite F-16 for 2004."

Vasiliev, S.V. (2005), "Potential for reductions of GHG emissions in the gas distribution sector" (Potentsial sokrashenia vibrosov parnikovih gazov v podotrasli gazoraspredelenia), presented at the Conference on the Dissemination of Methane Emission Control Experience in the Gas Industry, Higher School of Economics, Moscow, 11 March.

Venugopal, S. (2003), "Potential Methane Emissions Reductions and Carbon Offset Opportunities in Russia", TransCanada Pipelines Ltd., paper presented at the 3rd International Methane and Nitrous Oxide Mitigation Conference, 17-21 November, www.coalinfo.net.cn/coalbed/meeting/2203/papers/naturalgas/NG012.pdf.

Vernadsky Foundation (2004), "Kyoto Protocol in Russia: Prospect of JI Projects", presented at the Tenth Session of the Conference of the Parties (COP 9), Buenos Aires, 6-17 December, www.ieta.org/ieta/www/pages/getfile.php?docID=702.

Weiss, I. (2006), "Lukoil Targets Natural Gas in Ambitious Plan for Growth", *Oil Daily,* 25 May, Energy Intelligence Group.

World Bank (2004), "Options for Designing a Green Investment Scheme for Bulgaria", *PCFPlus research report*, http://carbonfinance.org/Router.cfm?Page=D ocLib&CatalogID=5928.

World Bank (2006), *Towards a World Free of Flares,* GGFR, Washington D.C.

WRI (World Resources Institute), WBCSD (World Business Council for Sustainable Development) (2004), *A Corporate Accounting and Reporting Standard,* GHG Protocol Initiative Web site.

WRI, WBCSD (2005), *The GHG Protocol for Project Accounting,* GHG Protocol Initiative Web site.

Wuppertal Institute (2005): Wuppertal Institute for Climate, Environment and Energy, Max-Planck-Institute for Chemistry (2005), "Greenhouse Gas Emissions from the Russian Natural Gas Export Pipeline System", Wuppertal, Mainz, www.wupperinst.org/download/1203-report-en.

WWF (World Wildlife Fund) (2005), "Emissions Trading is Possible, but only with a Certain Rationale" (Torgovat kvotami nado, no s umom), www.wwf.ru/news/article/2299.

Yukos (2003), *Exploration & Production, Regional Production Associations: OAO Tomskneft,* www.yukos.com/EP/Tomskneft.asp.

Zelinskiy, A.M. (2003), "Emissions Trading System in RAO UES Rossii Reform", presented at the Ninth Session of the Conference of the Parties (COP 9), Milan, 1-12 December.

Zhilin, O. (2004), "Gas Distribution Without Meters is Nonsense" (Razdavat gaz bez schechikov – eto absurd), *Gazprom journal* No. 2, pp. 22-23.

Zolotarevskiy, S.A., A.S. Osipov (2004), "Improvement of the Natural Gas Metering and Collection of Revenues from its Sales" (Sovershenstvovanie ucheta prirodnogo gaza i sobiraemosti sredstv za ego realizatsiu), *Territoria "NEFTEGAZ",* May, www.packo.ru/public/05.html.

USEFUL WEB SITES

EU-Russia Technology Centre (Round Table "Modern Technologies and Practices for Decreasing the Flaring of Associated Gas", Moscow, 19 April, 2005): www.technologycentre.org/content.php?topic=41.

GHG Protocol Initiative: www.ghgprotocol.org/templates/GHG5/layout.asp?MenuID=849.

MEDT (Ministry of Economic Development and Trade of the Russian Federation): www.economy.gov.ru/wps/portal.

Methane to Market Partnership: www.epa.gov/methanetomarkets/index.htm.

Natural Gas STAR (Recommended Technologies and Practices): www.epa.gov/gasstar/techprac.htm.

Natural Gas STAR (Methane Emissions from the Natural Gas Industry): www.epa.gov/gasstar/reports.htm.

GLOSSARY OF TERMS AND ABBREVIATIONS

AAU	Assigned Amount Unit, under the Kyoto Protocol
AIJ	Activities Implemented Jointly
BASREC	Baltic Sea Region Energy Co-operation
BAT	best-available technology
bbl	barrel
bcm	billion cubic meters
CAC	Central Asia Centre (pipeline)
CAPP	Canadian Association of Petroleum Producers
CDM	Clean Development Mechanism, under the Kyoto Protocol
CDM EB	CDM Executive Board
CENEf	Centre for Energy Efficiency, Russia
CEPA	Canadian Energy Pipeline Association
CH_4	methane
CIS	Commonwealth of Independent States: Republic of Armenia, Azerbaijan Republic, Republic of Belarus, Georgia, Kazakhstan, Kyrgyz Republic, Republic of Moldova, Russian Federation, Tajikistan, Turkmenistan, Ukraine, Republic of Uzbekistan.
CO_2	carbon dioxide
COP	Conference of the Parties to the United Nations Framework Convention on Climate Change (UNFCCC)
COP/MOP1	Conference of Parties to the UNFCCC serving as the Meeting of the Parties to the Kyoto Protocol
CSUGS	Compressor station for underground gas storage
DI&M	Direct Inspection and Maintenance
DN	digital number

DNA	Designated National Authority
DSM	demand-side management
DSMP	Defense Meteorological Satellite Programme
EIT	Economies in Transition: Russia, FSU (Former Soviet Union) and ECE (East & Central Europe)
ERU	Emission Reduction Units
ERUPT	Emission Reduction Unit Procurement Tender, Netherlands
ESCO	Energy Saving Company
ETS	Emissions Trading Scheme
EU	European Union
EU ETS	European Union Emissions Trading Scheme
FSU	Former Soviet Union
FTS	Federal Tariff Service of the Russian Federation
GDP	gross domestic product
GGFR	Global Gas Flaring Reduction Private-Public Partnership, World Bank Group
GHG	greenhouse gas
GIS	Green Investment Scheme
GOR	gas-to-oil ratio, expressed in m^3 of gas to m^3 of oil (m^3/m^3)
GPU	gas-powered compressor unit
GRI	Gas Research Institute, United States
GTL	gas-to-liquids
GWP	global warming potential
HFCs	hydrofluorocarbons (industrial greenhouse gases)
IEA	International Energy Agency
ICCC	Inter-Agency Commission of the Russian Federation on Climate Change

IETA	International Emissions Trading Association
IPCC	Intergovernmental Panel on Climate Change
IPIECA	International Petroleum Industry Environmental Conservation Association
IRR	internal rate of return (%)
JI	Joint Implementation, under the Kyoto Protocol
LNG	liquefied natural gas
LPG	liquefied petroleum gas
MEDT	Ministry of Economic Development and Trade of the Russian Federation
Mm³	million cubic metre
MtCO$_2$	million tonnes of carbon dioxide
MtCO$_2$e	million tonnes of carbon dioxide equivalent; this unit represents the equivalent CO$_2$ mass of greenhouse gases, reflecting their various global warming potential, usually computed over 100 years
MW	mega watt
NO$_2$	nitrous oxide
OECD	Organisation for Economic Co-operation and Development
OGP	International Association of Oil and Gas Producers
PCF	Prototype Carbon Fund, the World Bank Carbon Finance Unit
PDD	Project Design Document
PFCs	perfluorocarbons (industrial greenhouse gases)
PNNL	Pacific Northwest National Laboratory
PPP	purchasing power parity
PSA	production sharing agreement
R&D	research and development
RIIA	Royal Institute of International Affairs, now Chatham House

SBSTA	Subsidiary Body for Scientific and Technological Advice of the UNFCCC
SF$_6$	sulphur hexafluoride (industrial greenhouse gases)
SPE	Society of Petroleum Engineers
tCO$_2$	tonnes of carbon dioxide
TFC	total final consumption
TJ	tera joule (1×10^{12} joules)
toe	tonnes of oil equivalent
TPES	total primary energy supply
UES	United Electricity System of Russia (RAO UES)
UGSS	United Gas Supply System
UKPG	Unit of initial gas processing
µm	micrometre (1×10^{-6} m)
UNFCCC	United Nations Framework Convention on Climate Change
US	United States
USD	US dollar; USD 1 = EUR 0.81 (average exchange rate in 2005)
US EPA	United States Environment Protection Agency
VNIIGAZ	Russian Natural Gas Research Institute
WEO	World Energy Outlook
WRI	World Resources Institute
WSSD	World Summit on Sustainable Development
WTO	World Trade Organisation
WWF	World Wildlife Fund

The Online Bookshop

International Energy Agency

All IEA publications can be bought
online on the IEA Web site:

www.iea.org/books

You can also obtain PDFs of
all IEA books at 20% discount.

Books published before January 2005
- with the exception of the statistics publications -
can be downloaded in PDF, free of charge,
from the IEA Web site.

IEA BOOKS

Tel: +33 (0)1 40 57 66 90
Fax: +33 (0)1 40 57 67 75
E-mail: books@iea.org

International Energy Agency
9, rue de la Fédération
75739 Paris Cedex 15, France

IEA PUBLICATIONS, 9, rue de la Fédération, 75739 PARIS CEDEX 15
PRINTED IN FRANCE BY JOUVE
(61 2006 22 1P1) ISBN : 92-64-10986-2 - 2006